T0287275

Maker's Notebook
by the Staff of *Make:* magazine

Printed in U.S.A.

Published by Make: Books, an imprint of Make Community, LLC.
150 Todd Road, STE 200, Santa Rosa, CA 95407

Print History: April 2008: First Edition
February 2013: Second Edition
January 2021: Third Edition

Publisher: Dale Dougherty
Editors: Gareth Branwyn, Mike Senese
Copy Editors: Keith Hammond, Craig Couden, Caleb Kraft
Creative Director: Daniel Carter/Juliann Brown
Production Manager: Rob Bullington
Illustrator: Damien Scogin
Content: Gareth Branwyn, Terry Bronson, Paul Spinrad

ISBN-13: 9781680456639

Maker’s Notebook

TITLE

VOLUME

DATE

IN CASE OF LOSS, PLEASE RETURN TO:

NAME

ADDRESS

CITY

STATE

ZIP

LAND

MOBILE

EMAIL

WEBSITE

AS A REWARD: $

WELCOME TO YOUR MAKER'S NOTEBOOK

Creativity is inventing, experimenting, growing, taking risks, breaking rules, making mistakes and having fun. — Mary Lou Cook

When we started work on the Maker's Notebook in 2007, we set out to make a truly useful tool for makers. We wanted to create a project notebook that was extremely useful, attractive, and just plain fun to use. Having gone through two editions, many thousands of copies, and twelve years in the wild, we're proud of what we created. We've seen so many projects documented online that first came to life within these blue-grid pages. We've spotted Maker's Notebooks on the shelves of well-known makers like Adam Savage, and in the backgrounds of untold numbers of industrious YouTubers' workshops.

The Maker's Notebook has assisted in the creation of countless microcontroller projects, robot builds, costume and prop designs, 3D creations, craft projects, furniture and woodworking projects, house renovations, and much more. The notebook has been a home for your creative thinking and we're thrilled to keep that partnership alive with this third edition.

USING THE NOTEBOOK

We designed the Maker's Notebook so that it would appeal to a broad range of makers, from those using it in a professional setting, such as a lab or design studio, to more casual makers. It conforms to the basic guidelines for lab, engineering, and inventors' notebooks. These features include numbered, non-removable, quadrille graph paper (engineering-style, 1/8", acid-free), printed page numbers, places to name ideas/projects, and to sign, date, and witness each page.

In addition to the graph pages, we've included a section of updated reference material, with common weights and measures, project-oriented references (from 3D printing filament selection to sewing needle gauges), and a few bits of whimsical geekery just for fun (maker slang, anyone?).

MAKE: CHANGES

The Maker's Notebook was designed to be customized — hacked. The grid design on the front and back covers is begging for you to storyboard it with your own art, with stickers, to custom-design your own cover. When you do so, we want to see your handiwork! Upload your notebook cover (or any other parts you want to show off) to Instagram or Twitter and tag them with #makersnotebook.

Here are some other things you can do to customize your book:

- Create a keyword index. On the last blank page of your notebook, make a list of content that you want to organize, then make a little black mark on the outside edge of the page after each list item. As you add content to the notebook that comes under one of the subjects on your list, mark the outside edges of those pages with a black mark that corresponds to the placement of your list in the back. Now, you can find everything in the notebook under those categories by locating their black edge marks.

- Add additional horizontal or vertical pockets to the inside covers by using self-adhesive vinyl pockets (sold in office supply stores).

- If you use several Maker's Notebooks for different purposes (e.g., one for hardware hacking, one for home design/remodeling, one for business), you can use icons from the sticker sheet in the back to differentiate the books on their spines and covers.

- Clip a flat "bookmark pen" onto your notebook, so you always have a writing tool handy.

These are just a few ideas. If you come up with others, please share them online using the #makersnotebook tag.

The Maker's Notebook is a place to capture your creative thoughts and to assist your mind in designing the world around you. Since the launch of *Make:* magazine in 2005, we've dedicated ourselves to creating a thriving international community that lives to "Make, Hack, Build." But before all that, there's "Design." That's where the Maker's Notebook comes in. So, what are you waiting for?

— Your friends at *Make:*

Maker's Notebook

CONTENTS

CONTENTS

IDEA/PROJECT
DATE
1
NOTES/SIG
FROM PAGE
TO PAGE

IDEA/PROJECT

DATE

NOTES/SIG

FROM PAGE

TO PAGE

IDEA/PROJECT

DATE

NOTES/SIG

FROM PAGE

TO PAGE

IDEA/PROJECT

DATE

NOTES/SIG

FROM PAGE

TO PAGE

IDEA/PROJECT
DATE
5
NOTES/SIG
FROM PAGE
TO PAGE

IDEA/PROJECT

DATE

NOTES/SIG

FROM PAGE

TO PAGE

IDEA/PROJECT

DATE

NOTES/SIG

FROM PAGE

TO PAGE

IDEA/PROJECT

DATE

NOTES/SIG

FROM PAGE

TO PAGE

IDEA/PROJECT

DATE

NOTES/SIG

FROM PAGE

TO PAGE

IDEA/PROJECT

DATE

NOTES/SIG

FROM PAGE

TO PAGE

DEA/PROJECT

DATE

NOTES/SIG

FROM PAGE

TO PAGE

IDEA/PROJECT

DATE

NOTES/SIG

FROM PAGE

TO PAGE

EA/PROJECT
DATE
13
OTES/SIG
FROM PAGE
TO PAGE

IDEA/PROJECT

DATE

NOTES/SIG

FROM PAGE

TO PAGE

IDEA/PROJECT

DATE

NOTES/SIG

FROM PAGE

TO PAGE

IDEA/PROJECT

DATE

NOTES/SIG

FROM PAGE

TO PAGE

IDEA/PROJECT

DATE

NOTES/SIG

FROM PAGE

TO PAGE

IDEA/PROJECT

DATE

NOTES/SIG

FROM PAGE

TO PAGE

IDEA/PROJECT

DATE

NOTES/SIG

FROM PAGE

TO PAGE

IDEA/PROJECT

DATE

NOTES/SIG

FROM PAGE

TO PAGE

IDEA/PROJECT
DATE
NOTES/SIG
FROM PAGE
TO PAGE

IDEA/PROJECT

DATE

NOTES/SIG

FROM PAGE

TO PAGE

IDEA/PROJECT

DATE

NOTES/SIG

FROM PAGE

TO PAGE

IDEA/PROJECT | DATE

NOTES/SIG | FROM PAGE | TO PAGE

IDEA/PROJECT | DATE |

NOTES/SIG | FROM PAGE | TO PAGE

IDEA/PROJECT

DATE

NOTES/SIG

FROM PAGE

TO PAGE

EA/PROJECT | DATE |

OTES/SIG | FROM PAGE | TO PAGE

IDEA/PROJECT

DATE

NOTES/SIG

FROM PAGE

TO PAGE

IDEA/PROJECT

DATE

NOTES/SIG

FROM PAGE

TO PAGE

IDEA/PROJECT

DATE

NOTES/SIG

FROM PAGE

TO PAGE

IDEA/PROJECT

DATE

NOTES/SIG

FROM PAGE

TO PAGE

IDEA/PROJECT

DATE

NOTES/SIG

FROM PAGE

TO PAGE

IDEA/PROJECT	DATE	
NOTES/SIG	FROM PAGE	TO PAGE

IDEA/PROJECT

DATE

NOTES/SIG

FROM PAGE

TO PAGE

IDEA/PROJECT

DATE

NOTES/SIG

FROM PAGE

TO PAGE

IDEA/PROJECT

DATE

NOTES/SIG

FROM PAGE

TO PAGE

IDEA/PROJECT

DATE

NOTES/SIG

FROM PAGE

TO PAGE

IDEA/PROJECT

DATE

NOTES/SIG

FROM PAGE

TO PAGE

DEA/PROJECT
DATE
39
NOTES/SIG
FROM PAGE
TO PAGE

IDEA/PROJECT

DATE

NOTES/SIG

FROM PAGE

TO PAGE

IDEA/PROJECT

DATE

NOTES/SIG

FROM PAGE

TO PAGE

IDEA/PROJECT

DATE

NOTES/SIG

FROM PAGE

TO PAGE

IDEA/PROJECT

DATE

NOTES/SIG

FROM PAGE

TO PAGE

IDEA/PROJECT
DATE
NOTES/SIG
FROM PAGE
TO PAGE

IDEA/PROJECT
DATE
45
NOTES/SIG
FROM PAGE
TO PAGE

IDEA/PROJECT

DATE

NOTES/SIG

FROM PAGE

TO PAGE

IDEA/PROJECT

DATE

NOTES/SIG

FROM PAGE

TO PAGE

IDEA/PROJECT

DATE

NOTES/SIG

FROM PAGE

TO PAGE

IDEA/PROJECT

DATE

NOTES/SIG

FROM PAGE

TO PAGE

IDEA/PROJECT

DATE

NOTES/SIG

FROM PAGE

TO PAGE

IDEA/PROJECT

DATE

NOTES/SIG

FROM PAGE

TO PAGE

IDEA/PROJECT

DATE

NOTES/SIG

FROM PAGE

TO PAGE

IDEA/PROJECT

DATE

NOTES/SIG

FROM PAGE

TO PAGE

IDEA/PROJECT | DATE |

NOTES/SIG | FROM PAGE | TO PAGE

IDEA/PROJECT

DATE

NOTES/SIG

FROM PAGE

TO PAGE

IDEA/PROJECT

DATE

NOTES/SIG

FROM PAGE

TO PAGE

IDEA/PROJECT

DATE

NOTES/SIG

FROM PAGE

TO PAGE

IDEA/PROJECT

DATE

NOTES/SIG

FROM PAGE

TO PAGE

IDEA/PROJECT
DATE
NOTES/SIG
FROM PAGE
TO PAGE

IDEA/PROJECT
DATE
61
NOTES/SIG
FROM PAGE
TO PAGE

IDEA/PROJECT

DATE

NOTES/SIG

FROM PAGE

TO PAGE

DEA/PROJECT

DATE

NOTES/SIG

FROM PAGE

TO PAGE

IDEA/PROJECT

DATE

NOTES/SIG

FROM PAGE

TO PAGE

IDEA/PROJECT

DATE

NOTES/SIG

FROM PAGE

TO PAGE

IDEA/PROJECT

DATE

NOTES/SIG

FROM PAGE

TO PAGE

IDEA/PROJECT

DATE

NOTES/SIG

FROM PAGE

TO PAGE

68
IDEA/PROJECT
DATE
NOTES/SIG
FROM PAGE
TO PAGE

IDEA/PROJECT

DATE

NOTES/SIG

FROM PAGE

TO PAGE

IDEA/PROJECT

DATE

NOTES/SIG

FROM PAGE

TO PAGE

IDEA/PROJECT
DATE
71
NOTES/SIG
FROM PAGE
TO PAGE

IDEA/PROJECT

DATE

NOTES/SIG

FROM PAGE

TO PAGE

IDEA/PROJECT

DATE

NOTES/SIG

FROM PAGE

TO PAGE

IDEA/PROJECT
DATE
NOTES/SIG
FROM PAGE
TO PAGE

IDEA/PROJECT

DATE

NOTES/SIG

FROM PAGE

TO PAGE

DEA/PROJECT
DATE
77
OTES/SIG
FROM PAGE
TO PAGE

IDEA/PROJECT

DATE

NOTES/SIG

FROM PAGE

TO PAGE

IDEA/PROJECT

DATE

NOTES/SIG

FROM PAGE

TO PAGE

IDEA/PROJECT

DATE

NOTES/SIG

FROM PAGE

TO PAGE

IDEA/PROJECT
DATE
81
NOTES/SIG
FROM PAGE
TO PAGE

IDEA/PROJECT

DATE

NOTES/SIG

FROM PAGE

TO PAGE

IDEA/PROJECT

DATE

NOTES/SIG

FROM PAGE

TO PAGE

IDEA/PROJECT
DATE
NOTES/SIG
FROM PAGE
TO PAGE

IDEA/PROJECT
DATE
NOTES/SIG
FROM PAGE
TO PAGE

IDEA/PROJECT

DATE

NOTES/SIG

FROM PAGE

TO PAGE

IDEA/PROJECT
DATE
87
NOTES/SIG
FROM PAGE
TO PAGE

IDEA/PROJECT

DATE

NOTES/SIG

FROM PAGE

TO PAGE

IDEA/PROJECT
DATE
89
NOTES/SIG
FROM PAGE
TO PAGE

IDEA/PROJECT

DATE

NOTES/SIG

FROM PAGE

TO PAGE

IDEA/PROJECT

DATE

NOTES/SIG

FROM PAGE

TO PAGE

IDEA/PROJECT

DATE

NOTES/SIG

FROM PAGE

TO PAGE

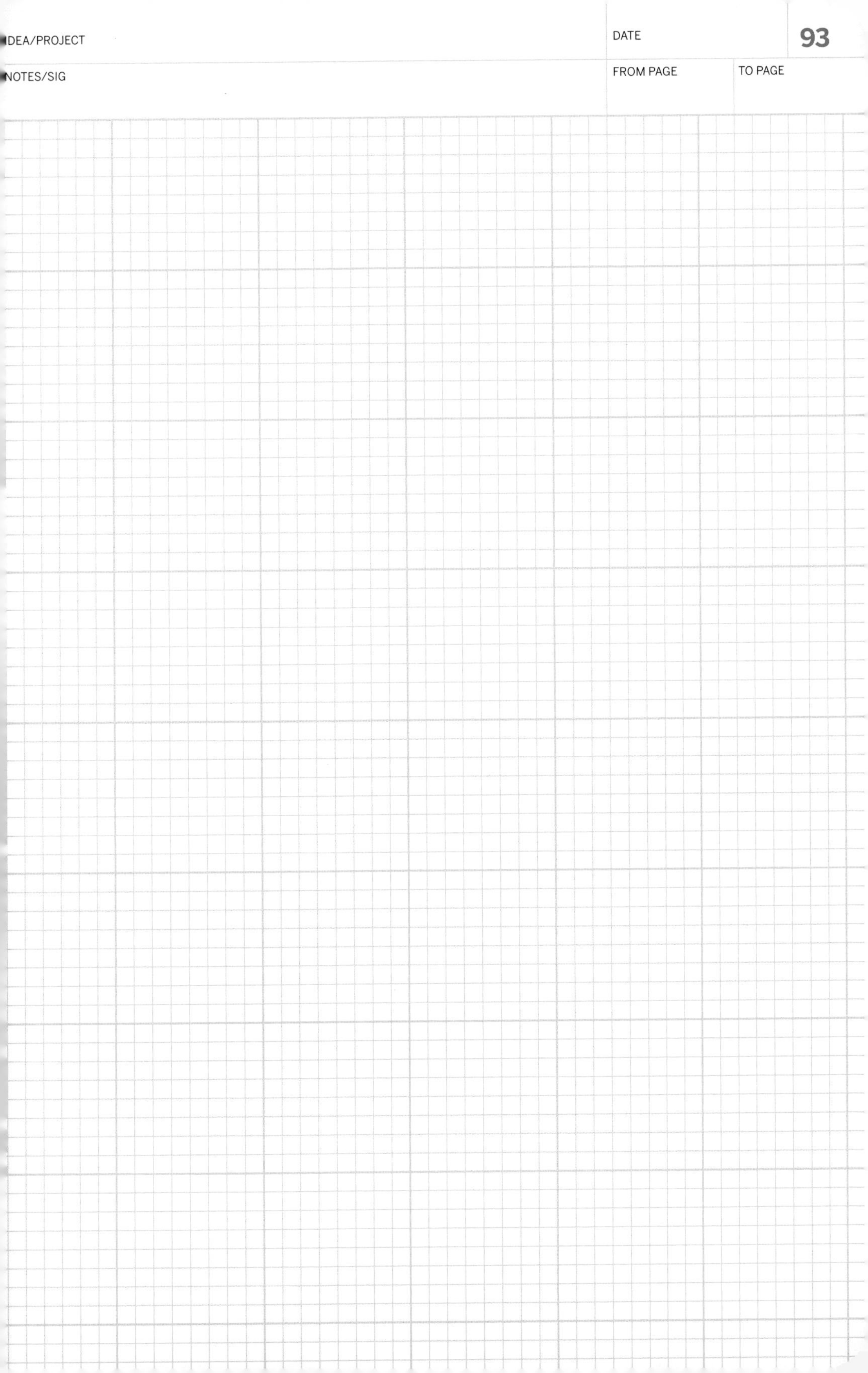
IDEA/PROJECT
DATE
93
NOTES/SIG
FROM PAGE
TO PAGE

IDEA/PROJECT

DATE

NOTES/SIG

FROM PAGE

TO PAGE

IDEA/PROJECT

DATE

NOTES/SIG

FROM PAGE

TO PAGE

IDEA/PROJECT

DATE

NOTES/SIG

FROM PAGE

TO PAGE

IDEA/PROJECT

DATE

NOTES/SIG

FROM PAGE

TO PAGE

IDEA/PROJECT

DATE

NOTES/SIG

FROM PAGE

TO PAGE

IDEA/PROJECT

DATE

NOTES/SIG

FROM PAGE

TO PAGE

IDEA/PROJECT

DATE

NOTES/SIG

FROM PAGE

TO PAGE

IDEA/PROJECT

DATE

NOTES/SIG

FROM PAGE

TO PAGE

IDEA/PROJECT

DATE

NOTES/SIG

FROM PAGE

TO PAGE

IDEA/PROJECT	DATE	
NOTES/SIG	FROM PAGE	TO PAGE

IDEA/PROJECT

DATE

NOTES/SIG

FROM PAGE

TO PAGE

IDEA/PROJECT

DATE

NOTES/SIG

FROM PAGE

TO PAGE

IDEA/PROJECT

DATE

NOTES/SIG

FROM PAGE

TO PAGE

IDEA/PROJECT

DATE

NOTES/SIG

FROM PAGE

TO PAGE

IDEA/PROJECT

DATE

NOTES/SIG

FROM PAGE

TO PAGE

IDEA/PROJECT
DATE
109
NOTES/SIG
FROM PAGE
TO PAGE

IDEA/PROJECT

DATE

NOTES/SIG

FROM PAGE

TO PAGE

IDEA/PROJECT

DATE

NOTES/SIG

FROM PAGE

TO PAGE

IDEA/PROJECT

DATE

NOTES/SIG

FROM PAGE

TO PAGE

IDEA/PROJECT

DATE

NOTES/SIG

FROM PAGE

TO PAGE

IDEA/PROJECT
DATE
NOTES/SIG
FROM PAGE
TO PAGE

IDEA/PROJECT

DATE

NOTES/SIG

FROM PAGE

TO PAGE

IDEA/PROJECT

DATE

NOTES/SIG

FROM PAGE

TO PAGE

IDEA/PROJECT

DATE

NOTES/SIG

FROM PAGE

TO PAGE

IDEA/PROJECT

DATE

NOTES/SIG

FROM PAGE

TO PAGE

IDEA/PROJECT

DATE

NOTES/SIG

FROM PAGE

TO PAGE

IDEA/PROJECT

DATE

NOTES/SIG

FROM PAGE

TO PAGE

IDEA/PROJECT

DATE

NOTES/SIG

FROM PAGE

TO PAGE

IDEA/PROJECT

DATE

NOTES/SIG

FROM PAGE

TO PAGE

IDEA/PROJECT

DATE

NOTES/SIG

FROM PAGE

TO PAGE

IDEA/PROJECT

DATE

NOTES/SIG

FROM PAGE

TO PAGE

IDEA/PROJECT

DATE

NOTES/SIG

FROM PAGE

TO PAGE

IDEA/PROJECT

DATE

NOTES/SIG

FROM PAGE

TO PAGE

IDEA/PROJECT

DATE

NOTES/SIG

FROM PAGE

TO PAGE

IDEA/PROJECT

DATE

NOTES/SIG

FROM PAGE

TO PAGE

IDEA/PROJECT

DATE

NOTES/SIG

FROM PAGE

TO PAGE

IDEA/PROJECT

DATE

NOTES/SIG

FROM PAGE

TO PAGE

IDEA/PROJECT

DATE

NOTES/SIG

FROM PAGE

TO PAGE

IDEA/PROJECT

DATE

NOTES/SIG

FROM PAGE

TO PAGE

IDEA/PROJECT

DATE

NOTES/SIG

FROM PAGE

TO PAGE

IDEA/PROJECT

DATE

NOTES/SIG

FROM PAGE

TO PAGE

IDEA/PROJECT

DATE

NOTES/SIG

FROM PAGE

TO PAGE

IDEA/PROJECT

DATE

NOTES/SIG

FROM PAGE

TO PAGE

IDEA/PROJECT

DATE

NOTES/SIG

FROM PAGE

TO PAGE

IDEA/PROJECT
DATE
NOTES/SIG
FROM PAGE
TO PAGE

IDEA/PROJECT

DATE

NOTES/SIG

FROM PAGE

TO PAGE

DEA/PROJECT

DATE

NOTES/SIG

FROM PAGE

TO PAGE

IDEA/PROJECT

DATE

NOTES/SIG

FROM PAGE

TO PAGE

IDEA/PROJECT

DATE

NOTES/SIG

FROM PAGE

TO PAGE

IDEA/PROJECT

DATE

NOTES/SIG

FROM PAGE

TO PAGE

Reference

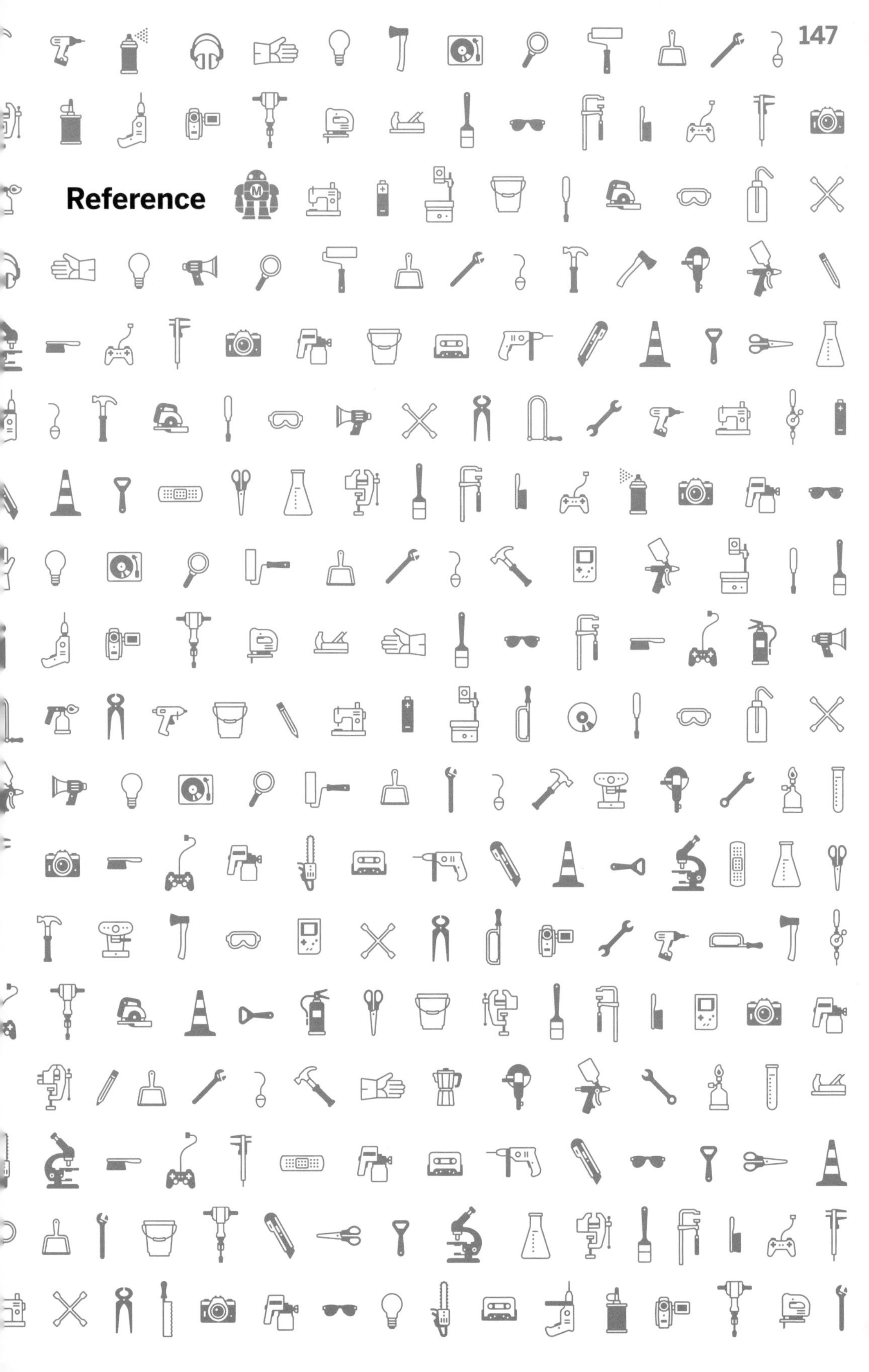

Common Weights and Measures

Length		
1 inch	1/36 yard	1/12 foot
1 yard	3 feet	
1 rod	5½ yards	
1 furlong	220 yards	40 rods
1 mile	1,760 yards	5,280 feet
1 fathom	6 feet	
1 nautical mile	6,076.1 feet	

Length (Metric System)	
1 millimeter	1/1000 meter
1 centimeter	1/100 meter
1 decimeter	1/10 meter
1 meter (basic unit of length)	
1 dekameter	10 meters
1 kilometer	1,000 meters

Area		
1 square inch	1/1296 square yard	1/144 square foot
1 square yard (basic unit of area)		
1 square rod	30¼ square yards	
1 acre	4,840 square yards	160 square rods

Volume (Dry & Liquid)		
1 cubic inch	1/46656 cubic yard	1/1728 cubic foot
1 cubic yard (basic unit of volume)		
1 U.S. fluid oz.	1/128 U.S. gallon	1/16 U.S. pint
1 pint	1/8 gallon	½ quart
1 U.S. gallon	231 cubic inches	
1 dry pint	1/64 bushel	½ dry quart
1 dry quart	1/32 bushel	1/8 peck
1 peck	¼ bushel	
1 U.S. bushel	2,150.4 cubic inches	

Weight (Avoirdupois)		
1 grain	1/7000 pound	1/437.5 ounce
1 dram	1/256 pound	1/16 ounce
1 ounce	1/16 pound	
1 pound	16 ounces	
1 short hundred wt.	100 pounds	
1 long hundred wt.	112 pounds	
1 short ton	2,000 pounds	
1 long ton	2,240 pounds	

Conversion Calculations		
To change	**To**	**Multiply by:**
centimeters	inches	0.3937
centimeters	feet	0.03281
cubic feet	cubic meters	0.0283
cubic meters	cubic feet	35.3145
cubic meters	cubic yards	1.3079
cubic yards	cubic meters	0.7646
degrees	radians	0.01745
feet	meters	0.3048
feet	miles (statute)	0.0001894
feet/second	miles/hour	0.6818
gallons (U.S.)	liters	3.7853
grams	ounces (avdp)	0.0353
grams	pounds	0.002205
horsepower	watts	745.7
horsepower	Btu/hour	2,547
hours	days	0.04167
inches	millimeters	25.4
inches	centimeters	2.54
kilograms	pounds	2.2046
kilometers	miles	0.6214
kilowatt-hour	Btu	3412
liters	gallons (U.S.)	0.2642
liters	pints (dry)	1.8162
liters	pints (liquid)	2.1134
liters	quarts (dry)	0.9081
liters	quarts (liquid)	1.0567
meters	feet	3.2808
meters	miles	0.0006214
meters	yards	1.0936
miles	kilometers	1.6093
miles	feet	5280
millimeters	inches	0.0394
ounces	pounds	0.0625
pounds (avdp)	kilograms	0.4536
pounds	ounces	16
quarts (dry)	liters	1.1012
quarts (liquid)	liters	0.9463
square feet	square meters	0.0929
square kilometers	square miles	0.3861
square meters	square feet	10.7639
square miles	square kilometers	2.59
square yards	square meters	0.8361
watts	Btu/hour	3.4121
watts	horsepower	0.001341
yards	meters	0.9144
yards	miles	0.0005682

Classic Design Tradeoffs

Time/Space, Compression/Quality, Cost/Time, Speed/Accuracy, Convenience/Privacy, Reliability/Maintenance Time, Simplicity/Flexibility, Security/Flexibility, Security/Freedom, Power/Efficiency, Scalability/Performance, Resolution/Overhead, Bandwidth/Latency.

Conversions

Fractions	Decimals
1/2	0.5000
1/3	≈ 0.3333
1/4	0.2500
1/5	0.2000
1/6	≈ 0.1667
1/7	≈ 0.1429
1/8	0.1250
1/9	≈ 0.1111
1/10	0.1000
1/11	≈ 0.0909
1/12	≈ 0.0833
1/16	0.0625
1/32	≈ 0.0313
1/64	≈ 0.0156
2/3	≈ 0.6667
2/5	0.4000
2/7	≈ 0.2857
2/9	≈ 0.2222
2/11	≈ 0.1818
3/4	0.7500
3/5	0.6000
3/7	≈ 0.4286
3/8	0.3750
3/10	0.3000
3/11	≈ 0.2727
4/5	0.8000
4/7	≈ 0.5714
4/9	≈ 0.4444
4/11	≈ 0.3636
5/6	≈ 0.8333
5/7	≈ 0.7143
5/8	0.6250
5/9	≈ 0.5556
5/11	≈ 0.4545
5/12	≈ 0.4167
6/7	≈ 0.8571
6/11	≈ 0.5455
7/8	0.8750
7/9	≈ 0.7778
7/10	0.7000
7/11	≈ 0.6364
7/12	≈ 0.5833
8/9	≈ 0.8889
8/11	≈ 0.7273
9/10	0.9000
9/11	≈ 0.8182
10/11	≈ 0.9091
11/12	≈ 0.9167

Length: Metric Conversions	
1 centimeter	0.39 inch
1 inch	2.54 centimeters
1 meter	39.37 inches
1 foot	0.305 meter
1 meter	3.28 feet
1 yard	0.914 meter
1 meter	1.094 yards
1 kilometer	0.62 mile
1 mile	1.609 kilometers

Volume: Metric Conversions	
1 cubic centimeter	0.06 cubic inch
1 cubic inch	16.4 cubic centimeters
1 cubic yard	0.765 cubic meter
1 cubic meter	1.3 cubic yards
1 milliliter	0.034 fluid ounce
1 fluid ounce	29.6 milliliters
1 U.S. quart	0.946 liter
1 liter	1.06 U.S. quarts
1 U.S. gallon	3.8 liters
1 liter	0.9 dry quart
1 dry quart	1.1 liters

	°Fahrenheit	°Celsius
Boiling point of water	212°	100°
Freezing point of water	32°	0°
Absolute zero	−459.6°	−273.1°

Caffeine

Substance	Ounces	Milligrams
Cocaine Energy Drink	8.4 oz	280 (33.3/oz)
Jolt Cola	23.5 oz	280 (11.9/oz)
SoBe No Fear Super Energy	16 oz	174 (10.9/oz)
Monster Energy	16 oz	160 (10/oz)
Rockstar	16 oz	160 (10/oz)
Full Throttle	16 oz	144 (9/oz)
Yerba Mate, traditional	6 oz	110
Coffee	8 oz	65–120
Espresso shot	1 oz	65–130
Foosh Energy Mint		100+/mint
Red Bull	8.3 oz	80 (9.16/oz)
Mountain Dew	12 oz	54
Tea, green or black	8 oz	14–61
Yerba Mate tea	8 oz	85
Diet Coke, Tab	12 oz	47
Dr Pepper	12 oz	36
Sunkist	12 oz	41
Pepsi	12 oz	38
Coca-Cola Classic	12 oz	35
Club Mate caffeine dose		20mg/100ml

Recycling

Number	Abbreviation	Name	Common Uses
1	PET	Polyethylene terephthalate	Disposable beverage and food containers as well as synthetic fibers
2	HDPE	High-density polyethylene	Reusable beverage, food, and petroleum containers
3	PVC	Polyvinyl chloride	Pipes, signs, clothing
4	LDPE	Low-density polyethylene	Work surfaces, flexible containers, containers for acids and other chemicals
5	PP	Polypropylene	Textiles, ropes, food containers
6	PS	Polystyrene	Laboratory equipment, polystyrene foams
7	OTHER	ABS, polycarbonate, and others	Bottles, CDs, Lego bricks

Archaic Chemical Names

Archaic Name	Description
Air	Generic term for a gas
Aqua fortis	Nitric acid
Aqua regia	Kingly water, a mixture of nitric and hydrochloric acids used to dissolve gold
Argentum	Silver
Barium white	Barium sulfate
Bitter salt	Magnesium sulfate
Black ash	Impure form of sodium carbonate
Blue vitriol	Copper sulfate pentahydrate
Brimstone	Sulfur
Carbolic acid	Phenol
Cream of tartar	Potassium bitartrate
Dragon's blood	Bright red resin from plant sources such as rattan palm
Ferrum	Iron
Green vitriol	Iron sulfate
Isinglass	Collagen obtained from fish, used in the clarification of fermented spirits
Jeweler's rouge	Iron oxide, used to polish metallic jewelry
Lampblack	Carbon black, a crude charcoal
Lime	Calcium oxide
Lime-water	A saturated calcium hydroxide solution
Muriatic acid	Hydrochloric acid
Potash	Potassium chloride, and other potassium-containing compounds
Prussic acid	Hydrogen cyanide
Quicklime	Calcium oxide
Quicksilver	Mercury
Saltpeter	Potassium nitrate
Soda ash	Sodium carbonate, also known as washing soda
Tin crystals	Stannous chloride
Venetian red	Ferric oxide
Vegetable alkali	Potassium carbonate
Verdigris	Copper acetate; a green pigment formed by applying acetic acid to copper
Vitriol	Sulfuric acid

Common English <-> 1337 Character Substitutions

A 4, B 8, E 3, I |, L 1, O 0, S $, T 7, Z 2

Mnemonic Devices

Memory aids for Makers.

MEASUREMENT

Device: Knuckles and dips.
For: Remembering months. With your fists together, start with the first knuckle and go across, saying the months for each knuckle and dip. A knuckle is 31 days, a dip 30 (except February 28/29).

Device: King Hector died mysteriously drinking chocolate milk.
For: Metric system length measurements (greatest to least): Kilo, Hecto, Deka, Meter, Deci, Centi, Milli

MATHEMATICS

Device: How I need a drink, alcoholic of course, after the heavy chapters involving geodesy.
For: The numbers of Pi (by counting the letters) to 14 places: 3.1415926535897

Device: The pinch goes to the smaller number.
For: Remembering less than (<) and greater than (>)

SCIENCE

Device: Kids prefer cheese over fried green spinach.
For: Order of taxonomies in biology: Kingdom, Phylum, Class, Order, Family, Genus, Species

Device: Environment is ABC.
For: Remembering the parts of an environment: Abiotic (non-living), Biotic (living), and Cultural (human-made)

Device: Anyone can make pretty high heeled shoes.
For: Classification of humans: Animalia, Chordata, Mammalia, Primates, Hominidae, Homo, Sapien

Device: Mercury viewed Earth's many aspects joyfully sitting under Neptune.
For: Remembering the order of planets (and Asteroid Belt) from the Sun: Mercury, Venus, Earth, Mars, Asteroid Belt, Jupiter, Saturn, Uranus, Neptune

Device: OIL RIG
For: Oxidation Is Loss (of electrons), Reduction Is Gain (of electrons)

TECHNOLOGY

Device: Roy G. Biv
For: The visible electromagnetic spectrum: Red, Orange, Yellow, Green, Blue, Indigo, Violet

Device: Raging Martians invaded Roy G. Biv using x-ray guns.
For: Remembering waves of the electromagnetic spectrum from longest to shortest: Radio, Microwave, Infrared, Visible, Ultraviolet, X-Ray, Gamma Ray

Device: Please do not throw sausage pizza away.
For: Order of layers in the Open System Interconnection (OSI) computer network protocol: Physical Layer, Data Link Layer, Network Layer, Transport Layer, Session Layer, Presentation Layer, Application Layer

Device: Black beetles running on your garden bring very good weather.
For: Order of resistor color bands: Black, Brown, Red, Orange, Yellow, Green, Blue, Violet, Gold, White

Device: Twinkle twinkle little star, Power equals I squared R.
For: $P = I^2 \times R$, that is, *Power* (in watts) = $\textit{Current}^2$ (in amperes) × *Resistance* (in ohms)

Device: Good models know how Dunkin Donuts can make μ not petite.
For: The ordering of common Greek size prefixes (from largest to smallest): Giga, Mega, Kilo, Hecto, Deka, Deci, Centi, Milli, Micro (μ), Nano, Pico

Device: I feel rather negatively about cats.
For: Remembering that cathode is negative. Used by Dave Hrynkiw's (of Solarbotics.com) high school physics teacher.

Common Radio Call Signs

A – Alpha	**N** – November
B – Bravo	**O** – Oscar
C – Charlie	**P** – Papa
D – Delta	**Q** – Quebec
E – Echo	**R** – Romeo
F – Foxtrot	**S** – Sierra
G – Golf	**T** – Tango
H – Hotel	**U** – Umbrella
I – India	**V** – Victor
J – Juliet	**W** – Whisky
K – Kilo	**X** – X-ray
L – Lima	**Y** – Yankee
M – Mike	**Z** – Zulu

Hello, World! In Various Languages

APL
```
'Hello, world!'
```

BASIC
```
100 PRINT "Hello, World!"
110 END
```

C++
```
#include <iostream>
int main()
{
  std::cout << "Hello, world!\n";
}
```

Java
```
public class HelloWorld
{
  public static void main(String args[])
  {
    System.out.println("Hello, World");
  }
}
```

Lisp
```
(print "Hello, World!")
```

Perl
```
print "Hello, World!\n";
```

PHP
```
<?php
  echo "Hello, World!\n";
?>
```

Python
```
print "Hello, World!"
```

Ruby
```
ruby -e ' puts "Hello, World!" '
```

sh
```
echo "Hello, world!"
```

TCL
```
puts "Hello, World!"
```

Sewing Machine Needle Sizes

	American	European
Lighter	8	60
	9	65
	10	70
	11	75
	12	80
	14	90
	16	100
	18	110
Heavier	19	120

Choose a size 8/60 needle for lightweight fabric, a 10/70 for medium-weight material, a 14/90 or 16/100 for heavy fabrics like jeans, upholstery, canvas, etc., and 18/110 or 19/120 for the heaviest fabrics.

Common Technical Abbreviations

A	ampere (also "amp")
AAC	Apple audio codec
ABS	acrylonitrile butadiene styrene
A/D	analog-to-digital
ADC	analog-to-digital conversion (or "ADC")
Ah	ampere-hour
ASCII	American Standard Code for Information Interchange
ASIC	application-specific integrated circuit
AWG	American Wire Gauge
BFO	beat frequency oscillator
BGA	ball grid array
BJT	bipolar junction transistor
BOM	bill of materials
BP	bandpass
BTU	British Thermal Unit
C	coulomb (also "common" and "collector")
CAP	capacitor
CCD	charge-coupled device
CDMA	code-division multiple access
CMOS	complementary metal-oxide semiconductor
CNC	computer numerical control
CTS	Clear To Send
CW	continuous wave
DARPA	Defense Advanced Research Projects Agency
D/A	digital-to-analog
DAC	digital-to-analog conversion (or converter)
DHCP	Dynamic Host Configuration Protocol
DIP	double in-line package
DoF	degrees of freedom (or "depth of field")
DPDT	double pole, double throw
DRM	digital rights management
DSP	digital signal processing
EEPROM	electrically erasable programmable read-only memory
EIA	Electronics Industries Alliance
EMF	electromotive force (or "electromagnetic field")
EMI	electromagnetic interference
EOM	end of message
ESD	electrostatic discharge
F	farad (also "frequency")
FET	field-effect transistor
FLED	flashing LED
FPGA	field-programmable gate arrays
FREQ	frequency
FrieNDA	friendly non-disclosure agreement
Gnd	Ground (voltage level)
H	henry
HF	high frequency
HV	high voltage
Hz	hertz
I	current (actually stands for "intensity")
IEEE	Institute of Electrical and Electronic Engineers
IrDA	Infrared Data Association
IRQ	interrupt request
ISO	International Organization for Standardization
J	joule
JFET	junction field-effect transistor
JTAG	Joint Test Action Group
kWh	kilowatt hour
LDR	light-dependent resistor
LED	light-emitting diode
LP	low-pass
mA	milliamperes
mcd	microcandela
MCU	microcontroller unit (also "uC" or "µC")
MOS	metal-oxide-semiconductor
MOSFET	metal-oxide-semiconductor field-effect transistor
MUX	multiplex (or "multi-user experience")
N	newton (unit of force)
NC	normally closed (also "no contact")
NiMH	nickel metal hydride
NDA	non-disclosure agreement
NIH	not invented here
NIMBY	not in my backyard
NIST	National Institute of Standards and Technology
NO	normally open
NPN	negative-positive-negative
N-s	newton-second (unit of impulse)
OpAmp	operational amplifier
OSC	oscillator
PCB	printed circuit board
PCM	pulse-code modulation
pF	picofarad
PNG	portable network graphics
PNP	positive-negative-positive
POT	potentiometer
PV	photovoltaic
PWM	pulse-width modulation
PZ	piezoelectric
QR	Quick Response code
Qtz	quartz
R	resistance
RC	resistor-capacitor
RFI	radio frequency interference
RSSI	received signal strength indication
RTS	request to send
Rx	receive
RXD	received exchange data
SAE	Society of Automobile Engineers International
SDK	software development kit
SEO	search engine optimization
SI	International System of Units
SIP	single in-line package
SMD	surface-mounted devices
SMT	surface-mounted technology
S/N	signal-to-noise ratio
SPST	single pole, single throw
SQL	Structured Query Language
S/S	stainless steel
SSH	Secure Shell
SWG	Standard Wire Gauge
T	tesla (unit of flux density)
TLA	three-letter acronym
TO	transistor outline package
TTL	transistor-transistor logic
Tx	transmit
TXD	transmit exchange data (see "RXD")
UART	universal asynchronous receiver-transmitter
UHF	ultra high frequency
UPS	uninterrupted power supply
USS	United States Standard
V	volt
Vcc	main supply voltage (power)
Vdd	secondary supply voltage (power)
Vee	negative supply voltage (power)
Vss	negative supply voltage (power)
VFO	variable-frequency oscillator
VOM	volt-ohm meter (or multimeter)
XMTR	transmitter
YMMV	your mileage may vary
Z	impedance

10 Useful Tips for Makers

3D Printing Parts from McMaster-Carr — Many parts in the online McMaster-Carr catalog have 3D STEP files (a common format for 3D designs) that you can download, convert to STL (a digital file format), load them into a CAD program, and prepare them for 3D printing. This way, you can print a critical part in plastic, test-fit it, and make sure that it's the part you actually want before buying it.

Find Treasures with *Grandfather's* Search Term — If you want to find some amazing tool and material treasures, try setting up a search string notification on sites like Craigslist and eBay for the word *grandfather's* (possessive). That way, things like "I'm cleaning out my grandfather's workshop" will get messaged to you. Also search: *estate sale tools*.

Give Old Tools Away — Want to make someone's day and feel good in the process? Give away tools you no longer use or need. Give them to a school, library, homeless shelter, or local makerspace. Or to a friend or neighbor who needs them. It's a great way to give back, make new friends, and lighten your load.

Use Acetate Overlays in Your Notebook — When figuring out project wiring or other parts of a project design that might be subject to change, tape sheets of acetate over your design and mark on that with a dry erase marker. That way, you can continue to change things around until you're confident you have the arrangement you need or want.

Organize Your Shop for First-Order Retrievability — Arrange your workspace so that more commonly-used tools and materials are closest to you. This can reduce time in finding and retrieving what you need. Conversely, less frequently-used tools are farther away. Put everything on casters so that you shop can be arranged to suit your project needs.

Paint Dries Half a Shade Darker — Colors often dry half a shade darker than they appear when wet. Making this mental calculation while choosing, mixing, and applying paint is important in getting the final colors you desire. Bonus tip: Don't try to lighten a dark color by adding a lighter color to it. Add the darker color to the lighter and bring that up to the shade you're after.

Power-Loss Freezer Indicator — To help determine whether your freezer thawed (or partially thawed) during a power outage, freeze a cup of water, place a quarter on top, and place it in the freezer. When power is restored, if the quarter is unmoved, the freezer is safe. If it's partially submerged, you had some thawing, but the freezer contents are likely still safe. If the quarter is at the bottom, you had a significant failure and the freezer's contents are likely spoiled. By the way, if you end up with a fridge or freezer full of spoiled food, check your homeowner's insurance policy. Spoiled food is usually covered (if not the result of flooding).

Reviving Dead Markers — Bring a dead, alcohol-based marker (e.g., Sharpie, Magic Marker) back to life simply by removing the nib (however you get inside your particular marker) and depositing a few drops of isopropyl alcohol onto the felt material. It's usually the solvent the ink is mixed with that dries out before the pigment does.

Super Glue and Baking Soda — If you add baking soda to cyanoacrylate (CA) glue, it produces an incredibly strong material that you can cut, carve, sand, and drill. You can use this in all sorts of repair applications.

Using Legos for Mold Boxes — Lego bricks and plates make for a perfect, reusable, and resizable mold box. Nearly every hobbyist and many pros who do molding and casting use them.

Taken from Gareth Branwyn's book *Make: Tips and Tales From the Workshop*.

Maker Slang

Here are a few bits of slang heard around the maker community. Slang says a lot about a culture, its obsessions, its sense of humor, its creativity, and the challenges it faces.

BOLT STRIPPERS — A pair of pliers. Said because using pliers on a bolt, when the proper wrench is not available, is a stripped fastener waiting to happen.

DEAD BUG MODE — An IC chip with its pins in the air. Dead bug mode is used in freeform soldering where you construct a circuit without a circuit board by soldering the components directly to each other and to the pins of the IC.

GREEBLE — Bits of raised detail that are added to the surface of an object to make it look more complex and visually interesting. Used in movie special effects, scale modeling, cosplay, etc. The term originated with the special effects team working on the first *Star Wars* movies.

KNOLLING — A technique of grouping objects that are alike and arranging them at right angles to one another. Used for organizing or photographing collections of objects. Named after the very angular furniture of Florence Knoll.

MAGIC SMOKE — The caustic smoke that's produced when an over-stressed circuit or component fails and burns, literally letting out a puff of smoke. The joke goes that the device works because it contains magic smoke. If the part fries, you've "let the magic smoke out" so it no longer functions. Powering a device on for the first time is sometimes referred to as a "smoke test." Also: *blue smoke*, *magic blue smoke*, and *letting the genie out*.

MAKER MATH — When you do a cost/benefit analysis on making something vs. buying it, determine that it's much more cost effective to buy it, but you make it anyway because it'll be fun, educational, and it will have been made by you.

MONKEY TIGHT / GORILLA TIGHT / KONGED — Monkey tight = tight enough, gorilla tight = too tight, Konged = immovable.

PERCUSSIVE MAINTENANCE — An old military term for whacking the daylights out of something to try and get it working again.

PERSUADER — Any hammer, sledge, axe, or other tool used to "encourage" material to move.

SHOP SHOWER — Spraying off the *shop glitter* (work dust) before leaving the workshop.

TRANSPORTER ACCIDENT — A 3D print that has failed, creating a tangled mess of plastic where your object was supposed to be. Also: *spaghetti*, *spaghetti monster*, or *angel hair pasta*.

UNOBTANIUM — Unusual, costly, or theoretical material that is impossible to acquire. The opposite of *obtainium*, which is any material that you can beg, borrow, or steal.

WAVE A DEAD CHICKEN — To basically resort to voodoo — any irrational act — in a last-ditch effort to get something to work.

Credits: Andrew Lewis, Bryce Lynch, Geoff Meston, Steve Roberts, Lenore Edman

Web References

JUICE BARS (CREATIVITY LINKS)
Brain Pickings (brainpickings.org)
Creative Bloq (creativebloq.com)
DesignBuzz (designbuzz.com)
Oblique Strategies (rtqe.net/ObliqueStrategies)
Notcot (notcot.org)

PROJECT BUILDING/DIY
Do It Yourself (doityourself.com)
Family Handyman (familyhandyman.com)
Github (github.com)
Hackaday (hackaday.com)
Hackster (hackster.io)
Instructables (instructables.com)
Make: Projects (makeprojects.com)
Music from Outer Space (musicfromouterspace.com)
Thingiverse (thingiverse.com)

TIPS AND TOOLS
Cool Tools (kk.org/cooltools)
Gareth's Tips, Tools, and Shop Tales (getrevue.co/profile/garethbranwyn)
Maker Update (makerprojectlab.com)
Shop Hacks (facebook.com/groups/shophacks)
ToolGuyd (toolguyd.com)
Toolmonger (toolmonger.com)
Wheelhouse (wheelhouse.substack.com)

TECHNICAL RESOURCES
Adafruit Learning System (learn.adafruit.com)
American Wire Guide Sizes (powerstream.com/Wire_Size.htm)
Bolt Depot (boltdepot.com/fastener-information)
Chip Directory (chipdir.nl)
Datasheet Archive (datasheetarchive.com)
Electronics Conversions Formulas & References (rfcafe.com/references/electrical.htm)
Sci.Electronics.Repair FAQ (repairfaq.org)
Skill Builders (makezine.com/tag/skill-builder)
This to That (Glue database) (thistothat.com)

MAKE: COMMUNITY
Make: Magazine and Blog (makezine.com)
Maker Camp (makercamp.com)
Make: Facebook (facebook.com/makemagazine)
Maker Faire (makerfaire.com)
Make: Projects (makeprojects.com)
Maker Shed (makershed.com)

SUPPLIERS
Adafruit Industries (adafruit.com)
BG Micro (bgmicro.com)
Digi-Key Electronics (digikey.com)
Evil Mad Scientist Laboratories (evilmadscientist.com)
Jameco Electronics (jameco.com)
Lee Valley (leevalley.com)
Maker Shed (makershed.com)
Matter Hackers (matterhackers.com)
Micro-Mark (micromark.com)
McMaster-Carr (mcmaster.com)
Mouser Electronics (mouser.com)
Newark (newark.com)
Solarbotics (solarbotics.com)
SparkFun Electronics (sparkfun.com)
TAP Plastics (tapplastics.com)

SURPLUS
All Electronics (allelectronics.com)
American Science and Surplus (sciplus.com)
Army Surplus Warehouse (armysurpluswarehouse.com)
Electronic Goldmine (goldmine-elec-products.com)
Herbach and Rademan (herbach.com)
Marlin P. Jones (mpja.com)

Robot References

ASIMOV'S THREE (PLUS) LAWS OF ROBOTICS
0. A robot may not injure humanity or, through inaction, allow humanity to come to harm.
1. A robot may not harm a human being or, through inaction, allow a human being to come to harm.
2. A robot must obey orders given to it by human beings, except where such orders would conflict with the Zeroth or First Law.
3. A robot must protect its own existence, as long as such protection does not conflict with the Zeroth, First, or Second Law.

TILDEN'S LAWS OF ROBOTICS
1. A robot must protect its existence at all costs.
2. A robot must obtain and maintain access to a power source.
3. A robot must continually search for better power sources.

Also known as:
1. Protect thy ass.
2. Feed thy ass.
3. Move thy ass to better real estate.

THE RODNEY BROOKS RESEARCH HEURISTIC
Look for what is so obvious to everyone else that it's no longer on their radar, and put it on yours. Seek to uncover assumptions so implicit, they're no longer being questioned. Question them. [Used by MIT AI Lab Director and iRobot founder Rodney Brooks.]

THE KENNY ROGERS RULE
When a project build turns frustrating, ugly; when the cursing starts. Step away. Take a break. It almost never fails. Corollary: The extent to which you don't want to stop is inversely proportional to the extent to which you need to.

Why "Kenny Rogers"? Cause "You've got to know when to hold 'em / Know when to fold 'em / Know when to walk away..." [Used by Gareth Branwyn in *Absolute Beginner's Guide to Building Robots*.]

RECIPE FOR BUILDING BEHAVIOR-BASED ROBOTS
BBR and similar types of bottom-up bots are built in layers, upon the successes of previous layers:

1. Do simple things first.
2. Learn to do them flawlessly.
3. Add new layers of activity over the results of simple tasks.
4. Don't change the simple things.
5. Make new layers work as flawlessly as the previous ones.
6. Repeat ad infinitum.

Dumpster Dives

So, where do dumpster divers go to score?
» For general finds, big box retailers, stores having closing sales, and apartment complexes. Also high-end neighborhoods the night before large item pickup day, and lightly damaged areas in newly declared disaster zones.
» For electronics, engineering schools and electronics stores are obvious destinations. Divers will frequently query IT folks about any offices doing widespread upgrades.
» Construction and demolition sites are where building supplies and materials are found. And manufacturing districts. For instance, glass cutters routinely discard small pieces of glass and mirror.

Divers like to scrounge at night and usually carry a stick and a a flashlight, or wear a headlamp.

Common Bonds SELECT THE RIGHT GLUE FOR YOUR MATERIALS

MATERIALS	Paper	Fabric	Felt	Leather	Rubber	Foam	Styrofoam	Plastic	Metal	Ceramic	Glass	Balsa	Cork	Wood
Wood	W	C/W	Sp/C	W/C/Ca	C/Ca	C	2K/H	L/C	2K/C/L	C/Ca	C/Ca	W	W	L/W
Cork	H/W	H/L	W	Ca/C	Ca/C	2K	W	L/Ca	C/Ca	L/Ca	Si	W	W	
Balsa	W	H/W	W	Ca/C	C/Ca	C	2K/H	L/Ca	2K/Ca	L/Ca	C/Ca	W		
Glass	A/W	A	A	A/Ca	Ca	Sp	2K/Sp	C/L	2K/C	2K/C/L	2K/L			
Ceramic	A/H	Ca/A	Ca/A	Ca/A/C	C/Ca	A	Ce/C	L/Ca/C	2K/C/L	Ce/Ca				
Metal	A/H	A	C	C/Ca	C/Ca	C	2K/H	2K/C	2K/C					
Plastic	H/Sp	Sp/C	Sp/C	Sp/Ca	C/Ca	Ca	Ca/C	L/Ca/2K						
Styrofoam	Sp/C	A/H	Sp	A	L	L/A	A/Sp							
Foam	Sp	Sp	Sp	C	C	Sp								
Rubber	Ca/C	A/C	C	Ca	Ca									
Leather	F/Sp	F	2K	C/F										
Felt	A/H	F/H	H/F											
Fabric	A/H	F/H												
Paper	A/W													

A = All-purpose-glue
F = Fabric glue
Sp = Spray adhesive
H = Hot glue
C = Contact adhesives
L = Construction adhesive (Liquid Nails, Loctite)
Ce = Ceramic glue
Si = Silicone
W = Wood glue
Ca = Cyanoacrylate (super glue)
2K = Two-component adhesive

Material Considerations for CNC

ABS — Acrylonitrile butadiene styrene is a *terpolymer*, meaning a combination of three polymers. ABS is a versatile, impact-resistant material that is easy to add color to and construct things with.

ALUMINUM 6061 — Aluminum is a lightweight, durable, and corrosion-resistant metal that is electrically conductive. The most commonly used, general-purpose alloy is 6061 aluminum, which offers a great blend of strength and machinability.

CARBON STEEL — As defined by AISI, carbon steel contains a carbon content up to 2.1% by weight. Lower carbon steels suitable for machining include C1010 and C1018.

HDPE — HDPE stands for high-density polyethylene. It's an inexpensive, lightweight, chemical-resistant, food safe plastic that has a high strength-to-density ratio, making it well suited for many applications.

TITANIUM — Titanium has a low density but high strength and can be alloyed with aluminum and iron among other elements. It has a low heat conductivity requiring lower cutting speeds in order to prolong tool life.

ACRYLIC — Acrylic is a transparent thermoplastic derived from natural gas, more brittle than polycarbonate but more scratch resistant and available in many colors and textures

BRASS — Brass is an alloy of primarily copper and zinc. It's often called *free-machining brass* because it's hard enough to hold it's shape but soft enough to machine easily.

DELRIN — Delrin is the brand name for acetal homopolymer resin, wich is a very hard, high-strength engineering plastic. It can withstand temperatures from −40°F to 248°F and mills easily.

STAINLESS 316 — Stainless steel is a steel alloy with a minimum of 10.5% chromium. The presence of chromium creates a thin microscopic layer preventing corrosion and rust. Stainless 316 is more resistant to acids than its 304 counterpart.

TOOL STEEL — Tool steel contains carbon content between 0.5% and 1.5%. The presence of carbides plays a dominant role in the quality of tool steel, and its resistance to abrasion makes it well suited for hand tools and machine dies.

From the Bantam Tools CNC Machinist Reference Charts (bantamtools.com/guides-and-workflows). Used with permission.

G-code Reference

Common commands for the most popular CNC (computer numerical control) program language.

From the Bantam Tools CNC Machinist Reference Charts (bantamtools.com/guides-and-workflows). Used with permission.

G-code	Parameters	Command
G0	axes	Rapid Traverse
G1	axes, F	Straight feed
G4	P	Dwell
G20		Inch unit mode
G21		Millimeter unit mode
G28.3	axes	Select absolute position
G53		Select absolute coordinates
G54–G59		Select coordinates system 1–6
G90		Absolute positioning mode
G91		Incremental positioning mode

G-code	Parameters	Command
M3	S	Spindle on CW
M4	S	Spindle on CCW
M5		Spindle off
M6		Tool change
M8		Coolant on
M9		Coolant off
F	Feed rate	Specify feed rate
S	RPM	Set spindle speed
N	Line number	Label G-code block
P	Seconds	Specify dwell time

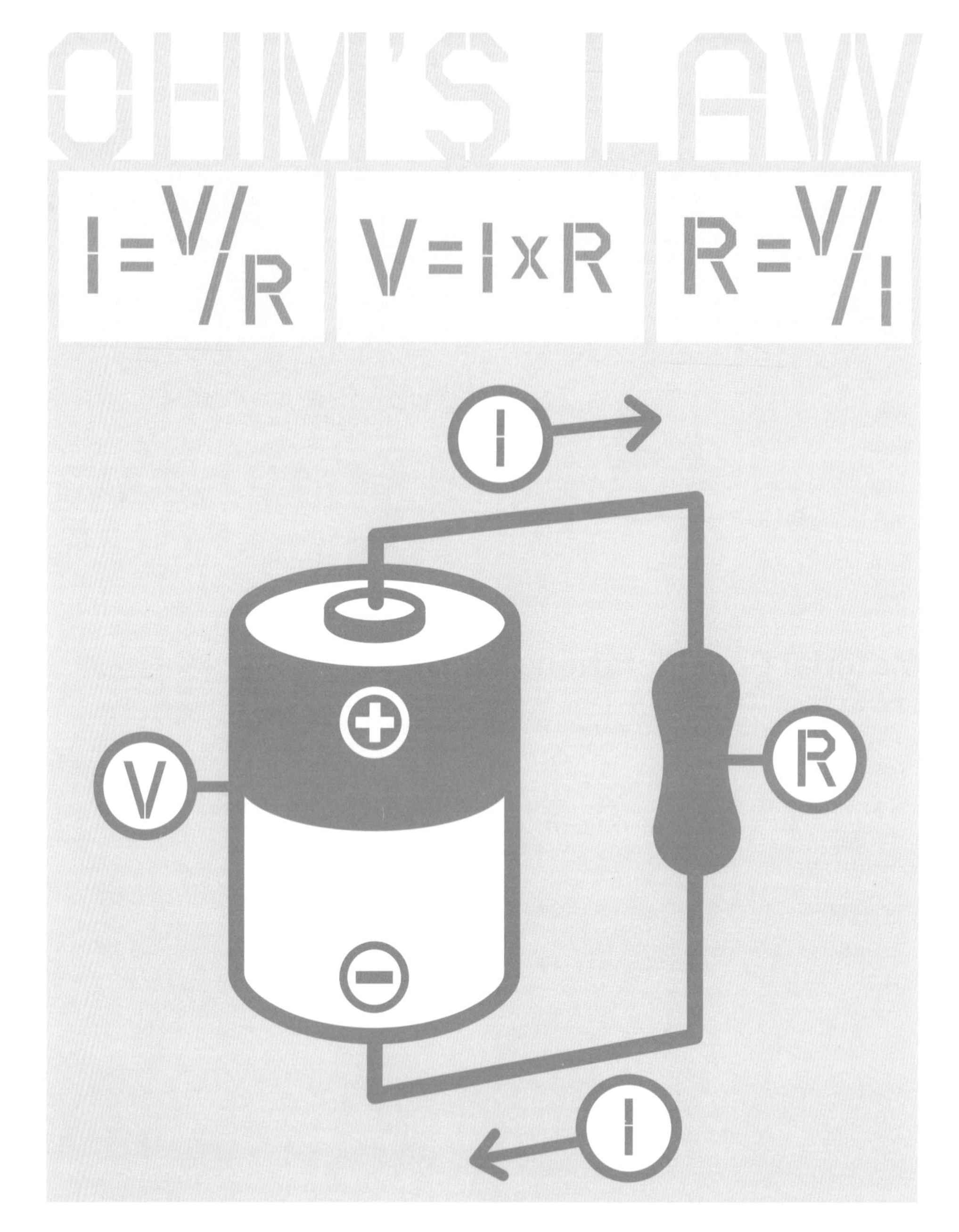

If a known current *(I)* flows through a resistor *(R)*, then voltage *(V)* can be calculated:
V = I × R

If you know the voltage *(V)* across a resistor *(R)*, then current *(I)* can be calculated:
I = V / R

If current *(I)* flows through a resistor, and there is a voltage across the resistor *(R)*, then resistance can be calculated: ***R = V / I***

(So, if *R* stands for resistance, and *V* for volts, what does the *I* for current stand for? Bet you didn't guess *intensity*.)

Basic Electronics Components and Function

Component	Symbol	Function	Measurement/ Classification	Notes
Fixed resistor		Restricts the flow of current in a circuit to a set value	Ohms (Ω)	Also measured in kΩ (1,000 ohms) and MΩ (1 million ohms)
Variable resistor		Restricts the flow of current in a circuit to a range of values	Ohms (Ω)	A popular type of variable resistor is called a potentiometer (or "pot") which uses a sliding contact or control dial.
Light-dependent resistor		A type of resistor that converts light to electrical resistance	Ohms (Ω) and lux (lx)	The resistance of an LDR decreases as the level of light increases.
Thermistor		A type of resistor that changes its resistance value in response to temperature changes	PTC (positive temperature coefficient) and NTC (negative temperature coefficient)	In a PTC thermistor, resistance increases with increasing temperature; in an NTC, resistance decreases with increasing temperature.
Capacitor	+	Temporarily stores current in a circuit and can also be used in filtering a signal	Farads (F)	Also measured μF (millionths). Can be either polarity-sensitive or non-polarity sensitive (fixed).
Diode	Anode Cathode	Restricts current flow to one direction only	A 1N-series numbering scheme, e.g., 1N34A/1N270 (germanium signal), IN914/1N4148 (silicon signal), and 1N4001–1N4007 (silicon 1A power rectifier)	There are a variety of diodes, including the most common (rectifying), Zener (which can conduct backwards at a set voltage), and Schottky diodes (useful in voltage clamping).
Light emitting diode (LED)	Anode Cathode	A type of diode that produces light when powered; used in everything from status indicator lights to flashlights and room lighting	Measured/classed by amps (mA), volts, candela (mcd), package size (e.g., 2mm, 3mm, 5mm), color, and shape	The cathode (–) side of the LED package is usually notched or flat. LEDs are available in light ranges from infrared to ultraviolet.
Transistor	NPN C B E PNP E B C	A type of semiconductor used to act as a switch in a circuit or to amplify a signal	Classed by structure (BJT, JFET, MOSFET), polarity, and power rating (low, med, high)	One of the most common types of transistor is the bi-polar junction transistor (BJT) which comes in two polarities (PNP and NPN) and has three pins: Emitter, Base, Collector.
Switch	L1 COM L2	A mechanical or electronic device used to open, close, or connect portions of an electrical circuit	Poles (numbers of circuits to control) and throws (numbers of circuit paths that are controllable)	Common switch types are designated as SPST (single pole, single throw), e.g., your average wall switch; SPDT, i.e., a switch that can switch between two circuits; and DPDT, i.e., a switch that can switch two circuits simultaneously.
Relay		A device that creates an electromagnetic field used to make (or break) an electrical contact (or contacts)	Pole and throws, and volts	A diagram showing the switch pins and the orientation of the electromagnet is frequently printed right on the relay package.
Integrated circuit	Vcc A B C Vdd IC NC Q NC Q̄ NC	A miniaturized electronic circuit that has been embedded in a substrate of semiconductor material	Type (analog, digital, hybrid), number of pins, function, chip family	IC pins are numbered, starting with pin 1 in the corner that's marked by a dimple and/or dot, then counting *counterclockwise* around the chip. (In this diagram, Vcc is pin 1.)
Battery	+ -	Used to store and deliver DC power to a circuit	Letter size (AAA, AA, C), ampere-hour (Ah), volts (V)	An array of electrochemical *cells* is called a *battery*. The polarity of most batteries is clearly marked and on most, the (+) side is a nipple or nub.
Motor	M	An electric motor is a device used to convert electromagnetic energy into mechanical motion	Volts, amps, RPM, and torque	Common types of DC motors include brushed and brushless, the *stepper motor*, and the *servomotor* (which is a control mechanism coupled with a DC gearmotor).
Transformer		A device for transferring electrical energy from one circuit to another by means of a magnetic field shared by the two circuits	Volts/amps, frequency, cooling type (air, oil, water), winding ratio (step-up, step-down), application and purpose (power supply, current stabilizer, rectifier, etc.)	
Lamp		A device that transforms electrical energy into light	Watts, candela, lumens, lux	On circuit diagrams, an "X" through the lamp symbol usually means that it is to be used as an indicator light.
Wire		A conductive material used to carry an electrical current; usually covered in an insulating material called a *jacket*	Measured in gauges, such as the American Wire Gauge (AWG)	In circuit diagrams, wires that connect are usually indicated by a dot on the junction of the wires.

Basic DMM Circuit Tests

BASIC DMM CIRCUIT TESTS

A Digital Multimeter (DMM) is a tool that every maker should own and know how to use. This chart outlines the basic tests that can be performed on AC and DC circuits (and individual components). These images show the general procedures, NOT the specifics for your multimeter. You must check the manual that came with your tool to find the right settings and test procedures.

VOLTAGE TESTS

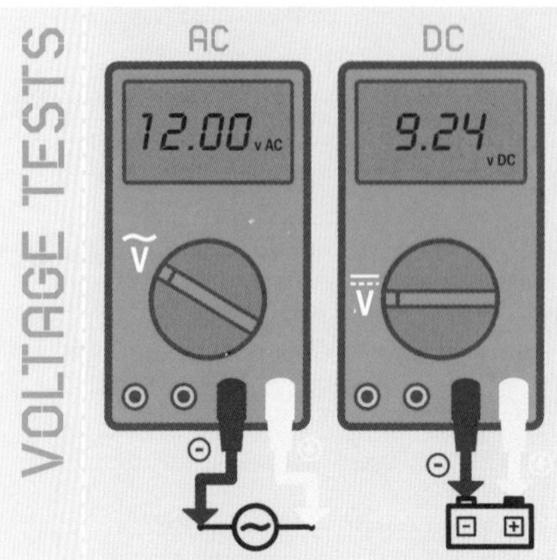

CAUTION! Do not test AC (i.e., house current) circuits unless you're already familiar with this type of power and how to handle it safely. You can hurt yourself, or worse.

RESISTANCE TEST

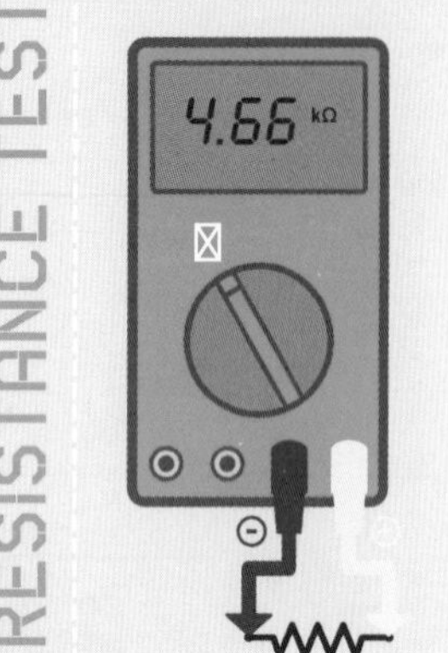

DIODE CHECK

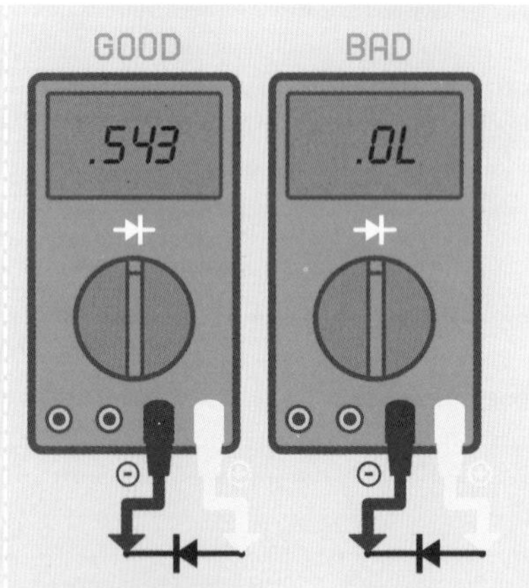

CONTINUITY TEST

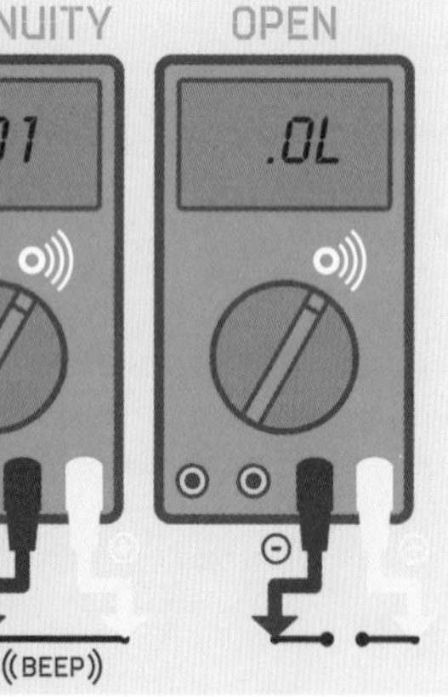

CURRENT TEST

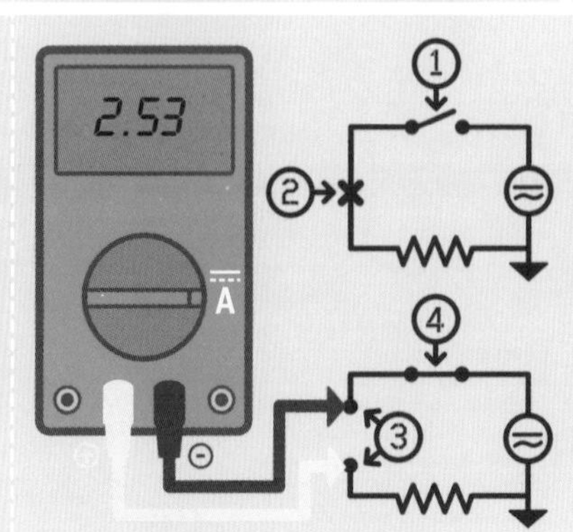

To do a current test, first, turn off power. Then you must "break" the circuit (cut or desolder part of it) and connect the DMM probes in series with the circuit (connecting at the points of the break). With probes connected, power up.

CAUTION! Make sure to put the red test probe in the A (amps) or mA (milliamps) jack on the DMM when testing current.

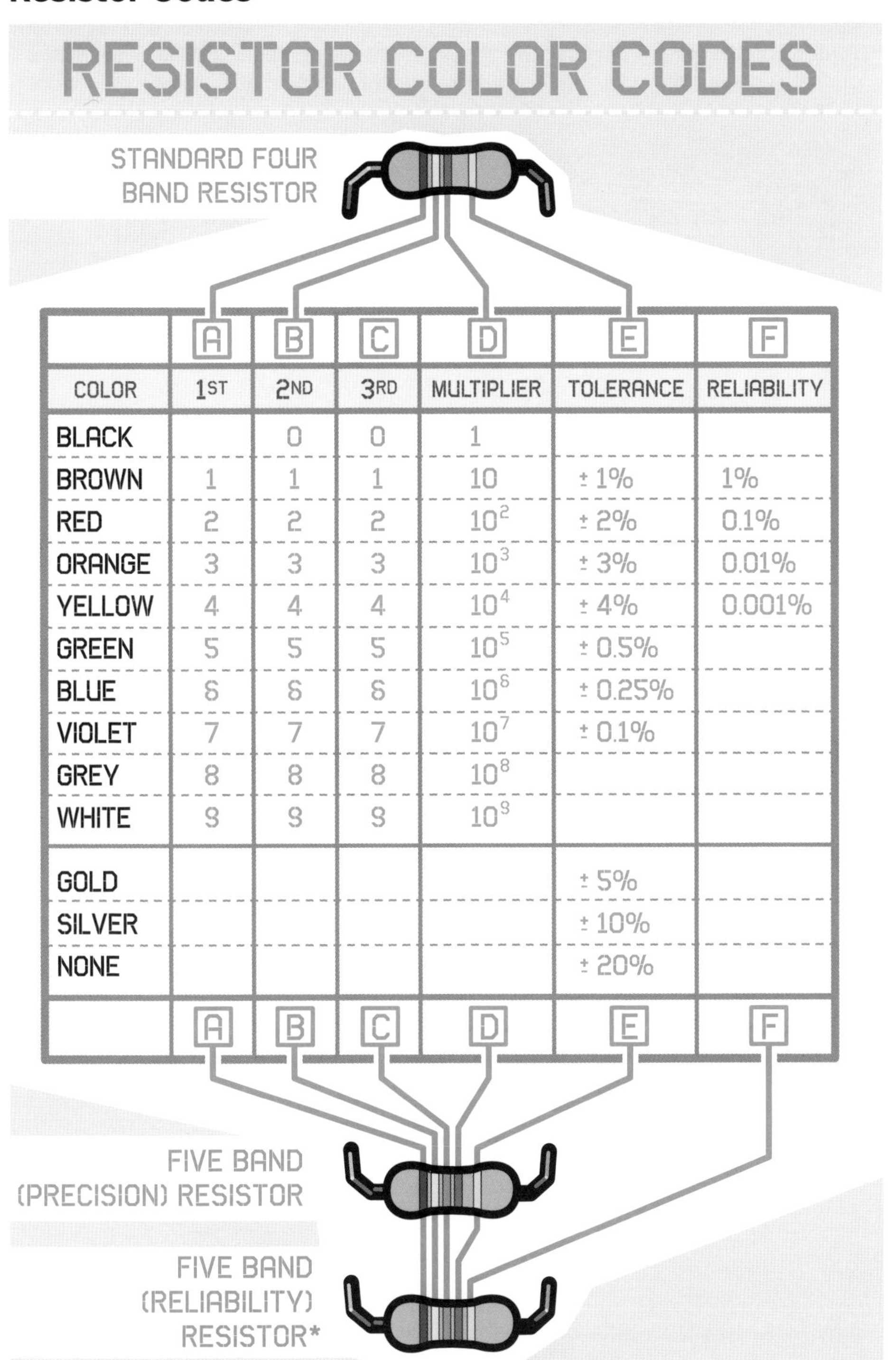

	A	B	C	D	E	F
COLOR	1ST	2ND	3RD	MULTIPLIER	TOLERANCE	RELIABILITY
BLACK		0	0	1		
BROWN	1	1	1	10	± 1%	1%
RED	2	2	2	10^2	± 2%	0.1%
ORANGE	3	3	3	10^3	± 3%	0.01%
YELLOW	4	4	4	10^4	± 4%	0.001%
GREEN	5	5	5	10^5	± 0.5%	
BLUE	6	6	6	10^6	± 0.25%	
VIOLET	7	7	7	10^7	± 0.1%	
GREY	8	8	8	10^8		
WHITE	9	9	9	10^9		
GOLD					± 5%	
SILVER					± 10%	
NONE					± 20%	
	A	B	C	D	E	F

Capacitors

Figuring out the values of capacitors from the strange numbers, symbols, and colors on them can be a bit of an arcane science. We didn't bother covering color-coded caps here, as those are not that common any longer. When in doubt: google.com.

ELECTROLYTIC CAPACITOR

These polarity-sensitive capacitors (caps) are the easiest to figure out. Their value, measured in farads (F) and their max. voltage are listed, in plain English, on the metal component "can." The negative (-) side is usually clearly marked and the negative lead is shorter.

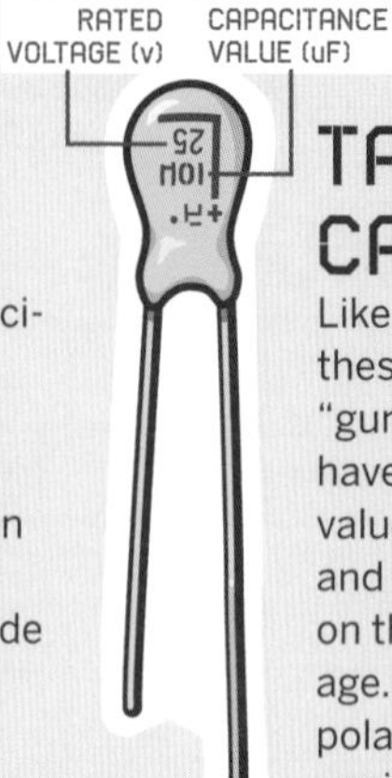

TANTALUM CAPACITOR

Like electrolytic caps, these brightly-colored "gumdrop" caps usually have the capacitance value, max. voltage rating, and polarity printed right on the component package. Tantalums are polarity-sensitive and the positive lead is longer than the negative.

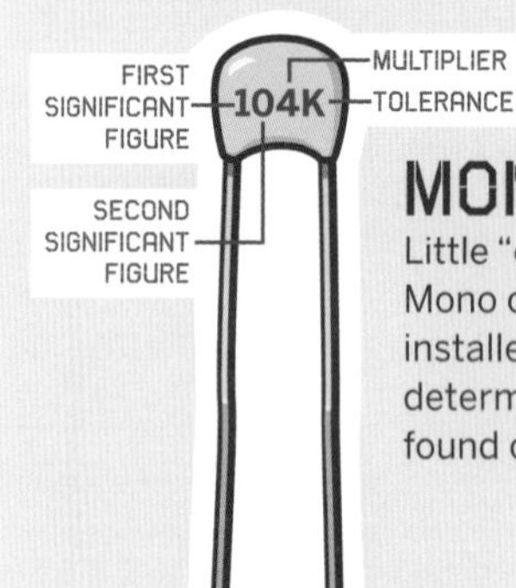

MONOLITHIC CAPACITOR

Little "chicklet" shaped caps in lots of fruit flavors (colors, anyway). Mono caps are non-polar, so the leads are equal length and can be installed in either direction. The chart below shows you how to determine their value using the three-number plus letter code found on most monolithic caps.

MONOLITHIC CAPACITOR CHART

VALUE (Fig 1 & 2)	MULTIPLIER	LETTER	TOLERANCE
0	1	B	± 0.1pF
1	10	C	± 0.25pF
2	10^2	D	± 0.5pF
3	10^3	F	± 1%
4	10^4	G	± 2%
5	10^5	H	± 3%
6	N/A	J	± 5%
7	N/A	K	± 10%
8	0.01	M	± 20%
9	0.1	Z	± 80%/-20%

CAPACITOR CODEBREAKER

COMMON CAPACITOR CODE	PICOFARAD (pF)	NANOFARAD (nF)	MICROFARAD (mF,uF or mfd)
102	1000	1 or 1n	0.001
152	1500	1.5 or 1n5	0.0015
222	2200	2.2 or 2n2	0.0022
332	3300	3.3 or 3n3	0.0033
472	4700	4.7 or 4n7	0.0047
682	6800	6.8 or 6n8	0.0068
103	10000	10 or 10n	0.01
153	15000	15 or 15n	0.015
223	22000	22 or 22n	0.022
333	33000	33 or 33n	0.033
473	47000	47 or 47n	0.047
683	68000	68 or 68n	0.068
104	100000	100 or 100n	0.1
154	150000	150 or 150n	0.15
224	220000	220 or 220n	0.22
334	330000	330 or 330n	0.33
474	470000	470 or 470n	0.47

The best way to have a good idea is to have lots of ideas. — Linus Pauling

LED Color Chart

Color Name	Wavelength (nm)	Fwd Voltage (Vf @ 20ma)	LED Dye Material
High Efficiency Red	640	2	GaAsP/GaP – Gallium Arsenic Phosphide/Gallium Phosphide
Super Red	634	2.2	InGaAIP – Indium Gallium Aluminum Phosphide
Red-Orange	623	2.2	InGaAIP – Indium Gallium Aluminum Phosphide
Orange	609	2.1	GaAsP/GaP – Gallium Arsenic Phosphide/Gallium Phosphide
Super Yellow	598	2.1	InGaAIP – Indium Gallium Aluminum Phosphide
Yellow	582	2.1	GaAsP/GaP – Gallium Arsenic Phosphide/Gallium Phosphide
Warm White	3000K	3.6	SiC/GaN – Silicon Carbide/Gallium Nitride
Pale White	6000K	3.6	SiC/GaN – Silicon Carbide/Gallium Nitride
Cool White	8000+K	3.6	SiC/GaN – Silicon Carbide/Gallium Nitride
Super Lime Yellow	575	2.4	InGaAIP – Indium Gallium Aluminum Phosphide
Super Lime Green	575	2	InGaAIP – Indium Gallium Aluminum Phosphide
High Efficiency Green	563	2.1	GaP/GaP – Gallium Phosphide/Gallium Phosphide
Super Pure Green	560	2.1	InGaAIP – Indium Gallium Aluminum Phosphide
Pure Green	555	2.1	GaP/GaP – Gallium Phosphide/Gallium Phosphide
Aqua Green	525	3.5	SiC/GaN – Silicon Carbide/Gallium Nitride
Blue Green	501	3.5	SiC/GaN – Silicon Carbide/Gallium Nitride
Super Blue	455	3.6	SiC/GaN – Silicon Carbide/Gallium Nitride
Ultra Blue	425	3.8	SiC/GaN – Silicon Carbide/Gallium Nitride

Source: LEDTronics.com. Used with permission.

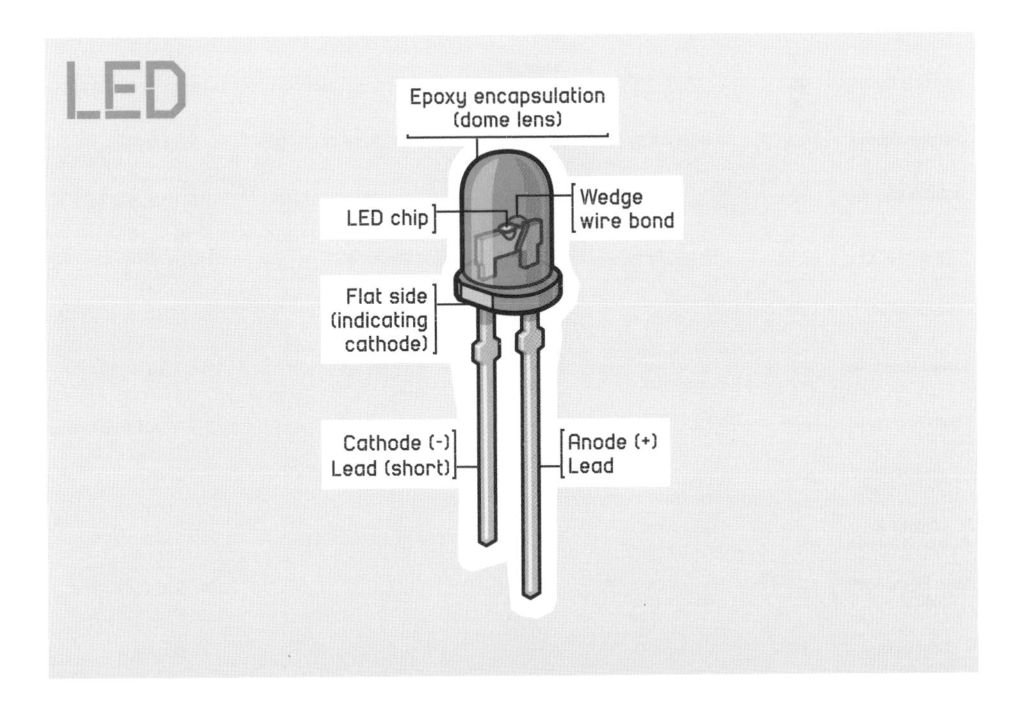

Microcontrollers (MCU)

BOARD NAME	PRICE	SOFTWARE	CLOCK SPEED	PROCESSOR	MEMORY
Adafruit Circuit Playground Express	$25	Arduino IDE, Circuit-Python, MakeCode, code.org CS Discoveries	48Mhz	32-bit ATSAMD21G18 Cortex-M0	256 KB flash, 32 KB RAM, 2 MB SPI flash
Adafruit Clue nRF52840 Express with Bluetooth LE	$40	Arduino IDE, CircuitPython	64MHz	32-bit Nordic nRF52840 Cortex-M4F	1MB flash, 256KB RAM
Adafruit Feather M4 Express	$23	Arduino IDE, Circuit-Python, MakeCode Maker	120MHz	32-bit ATSAMD51 Cortex-M4	512 KB flash, 192 KB RAM, 2MB SPI flash
Adafruit Metro M4 Express	$28	Arduino IDE, Circuit-Python, MakeCode	120MHz	32-bit ATSAMD51J19 Cortex-M4	512 KB flash, 192 KB RAM, 2 MB QSPI flash
Arduino Mega	$40	Arduino IDE	16MHz	8-bit ATmega2560	256KB flash, 8KB RAM, 4KB EEPROM
Arduino MKR WiFi 1010	$32	Arduino IDE	48MHz	32-bit SAMD21 Cortex-M0+	256KB flash, 32KB RAM
Arduino Nano Every	$13	Arduino IDE	20MHz	8-bit ATMega4809	48KB flash, 6KB RAM
Arduino Uno/Uno WiFi Rev2	Uno: $23 Uno WiFi: $45	Arduino IDE	16MHz	Uno: 8-bit ATMega328PU Uno WiFi: 8-bit ATmega4809	Uno: 32KB flash, 2KB RAM Uno WiFi: 32KB flash, 6KB RAM
BBC micro:bit	$15	MakeCode, Python	16MHz	32-bit Cortex-M0	16KB RAM
DFRobot Romeo V2	$35	Arduino IDE	16MHz	8-bit ATmega32u4	1MB RAM, 2MB flash, SD card slot
Espressif ESP32-S2 Saola-1	$8	Arduino IDE, Circuit-Python, ESP IDF, FreeRTOS	240MHz	32-bit ESP32-ST Xtensa LX7	4MB flash, 2MB PSRAM
Espruino Pico	$25	Espruino JavaScript Interpreter	84MHz	32-bit Cortex-M4	384KB flash, 96KB RAM
LilyPad Arduino USB	$25	Arduino IDE	16MHz	8-bit ATmega32u4	32KB flash
Meadow F7	$50	Meadow.OS (Runs apps written in the .NET developer framework)	216MHz	32-bit STM32F7 w/ESP32 Co-processor	32MB RAM, 32MB flash
Nordic Thingy:52	$40	Nordic Thingy	64MHz	32-bit nRF52832	512KB flash, 64KB RAM
OpenMV Cam H7	$65	MicroPython	480MHz	32-bit STM32H743VI Cortex-M7	1MB RAM, 2MB flash, SD card slot
Particle Boron	$57	Particle Device OS (FreeRTOS based)	64MHz w/ a 240MHz coprocessor	32-bit Cortex-M4F	1+4MB flash, 256KB RAM
Particle Photon	$19	Arduino IDE	100MHz	32-bit STM32F205 Cortex-M3	1MB flash
PJRC Teensy 4.1	$27	Arduino IDE with Teensyduino extension	600MHz	32-bit IMXRT1062 Cortex-M7	1MB RAM, 8MB flash
Pycom LoPy4	$40	MicroPython	160MHz	32-bit ESP32	4MB
Seeed Xiao	$5	Arduino IDE, CircuitPython	48MHz	32-bit ATSAMD21	256KB flash, 32KB SRAM
Sipeed Maixduino	$24	MaixPy IDE, Arduino IDE, OpenMV IDE, and PlatformIO IDE	400MHz	64-bit Sipeed M1 (RISC V)	16MB flash, 8MB SRAM
Sony Spresense	$65	NuttX emulating Arduino IDE	156MHz	32-bit Cortex-M4F × 6 cores	1.5MB SRAM, 8MB flash
SparkFun ESP8266 Thing	$16	Arduino IDE	80MHz	32-bit ESP8266	512KB flash
SparkFun Redboard Artemis	$20	Arduino IDE, Ambiq Apollo SDK	96MHz	32-bit Ambiq Apollo3 Cortex-M4F	384KB RAM, 1MB flash
Texas Instruments TM4C1294XL	$21	Energia, Code Composer, others	100MHz	32-bit Cortex-M4	1MB flash, 256KB RAM, 6KB EEPROM
TinyLily Mini	$10	Arduino IDE	8MHz	8-bit ATmega328P	32KB flash

Powered by Digi-Key Electronics (digikey.com). For more board specs go to makezine.com/boards

DIGITAL PINS	ANALOG PINS	RADIO	VIDEO	ETHERNET ON BOARD	INPUT VOLTAGE	BATTERY CONNECTION	OPERATING VOLTAGE	DIMENSIONS
8	8	–	–	–	5V	✓	3.3V	2" dia.
18	6	Bluetooth	1.3Ð 240×240 Color IPS TFT LCD display	–	3V–6V	✓	3.3V	2"×1.7"
15	6	–	–	–	3.3V	✓	3.3V	2"×0.9"
19	8 (2DAC)	–	–	–	7V–9V	–	3.3V	2.8"×2.1"
50+	7–12	–	–	–	6V–20V	–	5V	4"×2.1"
8–15	7	Wi-Fi, Bluetooth	–	–	3.7V–5V	✓	3.3V	2.5"×1"
14–22	8	–	–	–	7–21V	–	5V	1.8"×0.7"
14–20	1–5/6	Uno: No Uno Wifi: Wi-Fi, Bluetooth	–	–	Uno: 6V–20V Uno WiFi: 7–12V	–	5V	2.7"×2.1"
11–20	4–6	Bluetooth	–	–	1.8V–3.3V	–	3.3V	1.97"×1.57"
20	12 (6PWM)	Add-on modules available (Wi-Fi, Bluetooth)	–	–	5V	–	3.3V; Motor power 5V(2A)	3.5"×3.3"
43	20	Wi-Fi	–	–	3.3V	–	3.3V	2.22"×1.1"
22	9	–	Composite, VGA (through pins)	–	3.3V–16V	–	3.3V	1.3"×0.6"
5	4	–	–	–	2.7V–5.5V	–	3.3V	2" dia.
21	6 (+2)	Wi-Fi, Bluetooth	–	–	5V–12V	✓	3.3V	1.9"×0.9"
12–30	12–30	Bluetooth	–	–	5V	✓	3.3V	2.4"×2.4"
10	1 (10 PWM capable)	–	320×200 RGB camera	–	5V	–	3.3V	1.77"×1.41"
4–20	6	Celllar/BLE/NFC	–	–	USB 4.5V–5V, battery 3.6V–4.2V	✓	3.3V	2"×0.9"
11–20	7–12	Wi-Fi	–	–	3.6V–5.5V	–	3.3V	1.44"×0.8"
55	35 PWM, 18 ADC	–	–	✓	3.6V–5.5V	–	3.3V	2.4"×0.7"
5	18ADC, 2DAC, 13PWM	Wi-Fi, Bluetooth, Lora, Sigfox	–	–	3V–5.5V	–	3.3V	2.16"×0.7"
11	11	–	–	–	5V	–	3.3V	0.79"×0.69"
48	6	Wi-Fi, Bluetooth	LCD	–	1.8V–3.3V	–	3.3V	2.7"×2.1"
17; extension:14	2ADC; extension: 6 PWM, 6 ADC	–	–	–	5V	–	1.8V	1.96"×0.81"
7	1	Wi-Fi	–	–	3.3V–6V	–	3.3V	2.18"×1.02"
21	6 ADC (2V max), all PWM	Bluetooth	–	–	5V	–	3.3V	2.7"×2.1"
21–50	13+	–	–	✓	5V	–	3.3V	4.9"×2.2"
1–10	4–6	–	–	–	2.7V–5.5V	–	3V	0.55" dia.

SINGLE-BOARD COMPUTERS (SBC)

BOARD NAME	PRICE	SOFTWARE	CLOCK SPEED	PROCESSOR	MEMORY
Arduino Nano Every	$13	Arduino IDE	20MHz	8-bit ATMega4809	48KB flash, 6KB RAM
Asus Tinker Board S	$85	Debian Linux (Linaro), Android 6 & 7	1.8GHz	32-bit Rockchip RK3288	2GB dual channel DDR3
Asus Tinker Edge R	$250	Debian 9, Android 9	1.8GHz (Dual Core) + 1.4GHz (Quad Core)	64-bit Rockchip RK3399Pro (dual-core Cortex-A72 @ 1.8 GHz + quad-core Cortex-A53 @ 1.4 GHz)	4GB dual channel LPDDR4 (system) + 2GB LPDDR3 (NPU)
Asus Tinker Edge T	$160	Mendel Linux	1.5GHz	64-bit NXP i.MX 8M	1GB LPDDR4
Banana Pi M2 Berry	$36	Linux	1GHz	32-bit quad-core Cortex-A7 V40	1GB DDR3 SDRAM
BeagleBoard PocketBeagle	$25	Linux	1GHz	32-bit Cortex-A8	512MB DDR3
BeagleBoard-X15	$270	Linux	1GHz	32-bit AM5728 Cortex-A15	4GB 8-bit eMMC flash
BeagleBone AI	$99	Debian GNU/Linux	1.5GHz	32-bit Sitara AM5729 Cortex-A15, dual 32-bit Cortex-M4 coprocessors; SGX544 GPU	1GB RAM, 16GB flash
BeagleBone Black	$55	Linux	1GHz	32-bit AM335X Cortex-A8	4GB eMMC flash
BeagleBone Blue	$82	Debian Linux with Cloud9 IDE and libroboticscape	1GHz	32-bit Cortex-A8, TI Programmable Real-time Units	512MB RAM, 4GB eMMC flash
DFRobot LattePanda v1	$89	Windows 10	1.92GHz	64-bit quad-core Intel Z8350	2GB RAM 32GB flash
Google Coral Ed-geTPU Dev Board	$150	Mendel Linux, Android	1.3GHz	64-bit NXP i.MX8MQ	1GB
mangOH Yellow	$131	Linux, Legato	1.2GHz	32-bit Cortex-A7	4GB flash, 2GB DDR
Nvidia Jetson AGX Xavier Dev Kit	$699	Linux-based JetPack SDK	2.26GHz CPU, 1.37GHz GPU	64-bit 512-core Volta GPU with 64 Tensor cores; 64-bit 8-core ARM v8.2 CPU	32GB 256-bit LPDDR4
Nvidia Jetson Nano Dev Kit	$99	Linux-based JetPack SDK	1.43GHz CPU, 921MHz GPU	64-bit 128-CUDA-core Maxwell GPU; quad-core Cortex-A57 CPU	4GB 64-bit LPDDR4
Nvidia Jetson TX2 Dev Kit	$399	Linux-based JetPack SDK	2GHz CPU, 1.3GHz GPU	64-bit 256-core Pascal GPU; 64-bit Denver 1.5 and Cortex-A57 CPU	8GB L128-bit DDR4 LPDDR4
Nvidia Jetson Xavier NX Dev Kit	$399	Linux-based JetPack SDK	1.9GHz CPU, 1.1GHz GPU	64-bit Volta GPU with 384-CUDA-core and 48-Tensor-core GPU; 64-bit 6-core Carmel ARM CPU	8GB 128-bit LPDDR4x
Odroid-XU4	$59	Linux	2GHz	32-bit Samsung Exynos5422/32-bit octa-core Cortex-A15	2GB
Onion Omega2	$5	Linux	580MHz	32-bit MIPS	128MB
Qualcomm DragonBoard 410c	$75	Android, Linux, Win 10 IoT	1.2GHz	64-bit Snapdragon 410	1GB LPDDR3 533MHz, 8GB flash
Raspberry Pi 3, Model A+	$25	Raspbian Linux	1.4GHz	64-bit Broadcom BCM2837, quad-core A53 ARMv8	512MB
Raspberry Pi 4, Model B	$35–$75	Raspbian Linux, Other	1.5GHz	64-bit Broadcom BCM2711, quad-core	2, 4, or 8GB SDRAM
Raspberry Pi Zero W	$10	Rasbian Linux	1GHz	32-bit Broadcom ARMv6	microSD
Rock Pi 4	$49–$75	Android 7&9, Debian, Ubuntu, Other	1.8GHz + 1.4GHz	64-bit dual Cortex-A72, Quad Cortex-A53	1, 2, or 4GB
Seeed Odyssey	$188	Linux, Windows 10	1.5GHz	64-bit Intel Celeron J4105; 32-bit ATSAMD21 Cortex-M0+ coprocessor	8GB LPDDR4
UDOO Neo Full	$65	Linux	1GHz	32-bit Freescale i.MX 6SoloX ARM Cortex-A9	microSD
VoCore 2	$18	Linux	100MHz	16-bit MT7628AN, 580MHz, MIPS 24K	128MB, DDR2, 166MHz

Powered by Digi-Key Electronics (digikey.com). For more board specs go to makezine.com/boards

DIGITAL PINS	ANALOG PINS	RADIO	VIDEO	ETHERNET ON BOARD	INPUT VOLTAGE	BATTERY CONNECTION	OPERATING VOLTAGE	DIMENSIONS
14–22	8	–	–	–	7–21V	–	5V	1.8"×0.7"
26	3 PWM	Wi-Fi, Bluetooth	HDMI	✓	5V	–	5V	3.37"×2.125"
28	3 PWM	Wi-Fi, Bluetooth, Mini PCIe slot (for 4G/LTE)	HDMI, USB-C, MIPI DSI	✓	12V–19V	–	5V	3.9"×2.8"
28	3 PWM	Wi-Fi, Bluetooth	HDMI, MIPI DSI	✓	12V–19V	–	5V	3.37"×2.13"
26	–	Wi-Fi, Bluetooth	HDMI	✓	5V	–	3.3V	3.6"×2.4"
40	8ADC (6 at 1.8V, 2 at 3.3V)	–	–	–	5V	–	3.3V	2.2"×1.4"
50+	–	–	HDMI	✓	12V	–	3.3V	4"×4.2"
72	7	Wi-Fi, Bluetooth	Micro-HDMI	✓	5V	–	3.3V	3.4"×2.1"
50+	7–12	Wi-Fi	Micro-HDMI	✓	5V	–	1.8V–3.3V	3.4"×2.1"
1–7	4	Wi-Fi, Bluetooth	–	–	9V–18V	✓	1.8V–7.4V	3.4"×2.1"
12	6	Wi-Fi, Bluetooth	HDMI	✓	5V	–	5V	2.75"×3.42"
1–27	12–30	Wi-Fi, Bluetooth	HDMI	✓	5V	–	5V	3.3"×2.2"
6	2 ADC	Wi-Fi, Bluetooth, Cellular (NB-IOT, cat-m1, 2G, 3G, LTE cat-1, cat-4), NFC, GPS	–	–	4.75V–6V	✓	3.3V	1.65"×2.56"
28	PWM support	Wi-Fi	HDMI, DP	✓	9–20V	–	9V–20V; 5V	4.13"×4.13"× 2.56"
28	PWM support	–	HDMI, DP	✓	5V	–	5V	3.95"×3.15"× 1.14"
40+	PWM support	Wi-Fi, Bluetooth	HDMI, DP	✓	5.5V–19V	–	5.5V–19.6V	6.7"×6.7"× 1.97"
30	PWM support	Wi-Fi, Bluetooth	HDMI, DP	✓	9–20V	–	5V	4.06"×3.56"× 1.22"
25	–	–	HDMI	–	5V	–	1.8V	3.3"×2.3"
18	–	Wi-Fi	–	–	3.3V	–	3.3V	1.1"×1.7"
12	–	Wi-Fi, Bluetooth	HDMI	–	6.5V–18V	–	1.8V	2.12"×3.35"
24	–	Wi-Fi, Bluetooth	HDMI	–	5V	–	3.3V	2.6"×2.2"
29	Software-driven PWM	Wi-Fi, Bluetooth	Micro-HDM	✓	5V	–	3.3V	3.4"×2.2"
21–50	–	Wi-Fi, Bluetooth	Micro-HDMI	–	5V	–	3.3V	1.18"×2.56"
28	2	Wi-Fi, Bluetooth	HDMI	✓	6V–28V	–	3.3V	3.37"×2.22"
29	6	Wi-Fi, Bluetooth	HDMI	✓	12–19V	–	12V	4.33"×4.33"
21–50	4–6	Wi-Fi, Bluetooth	Micro-HDMI	✓	6V–15V	–	3.3V	3.5"×2.3"
21–50	4–6	Wi-Fi	–	✓	3.6V–6V	–	3.3V	1"×1"

3D Printer Filament Types

Material	Extruder Temp.	Bed Temp.	Adhesion Type	Notes
PLA – Polylactic Acid	205±15°C	35±15°C	Blue Painter's Tape or PVA glue	The most common and broadly useful printing material. Stiff but brittle. Odorless and low-warp, and does not require a heated bed. Made from annually renewable resources (cornstarch) and requires less energy than petroleum-based plastics.
ABS – Acrylonitrile Butadiene Styrene	230±10°C	90±10°C	Kapton Tape/Hairspray	Best for durable parts that need to withstand higher temperatures. Less brittle and more ductile than PLA. Can be post-processed with acetone for a glossy finish. Contracts when cooled, which can lead to warped parts; heated printing surface is recommended.
Nylon – Polyamide	255±15°C	60±10°C	PVA-Based Glue	Incredibly strong, durable, and versatile. Flexible when thin. Works well for living hinges and functional parts. Prints bright natural white with translucent surface. Can absorb color post-printing. Extremely moisture-sensitive; keep dry during storage and prior to printing.
PET (PETG, PETT) – Polyethylene Terephthalate	245±10°C	60±10°C	Blue Painter's Tape	Industrial strength filament. Much stronger than PLA, but, unlike ABS, barely warps and produces no odors or fumes when printed. Not biodegradable. Known for its clarity and is very good at bridging.
TPU – Thermoplastic Polyurethane	250±10°C	50±10°C	Blue Painter's Tape	Flexible filament with shore hardness of 95A. Creates rubbery, elastic, and impact resistant parts. Best used for stoppers, belts, caps, phone cases, bumpers and more. The less infill you use, the more flexible your finished print will be.
PVA – Polyvinyl Alcohol	180±20°C	45±10°C	Blue Painter's Tape	Often used with dual extruders, one printing a primary material (such as ABS or PLA) and the other printing PVA to provide support for overhanging features. Dissolves when exposed to water. MUST be kept dry prior to use.
HIPS – High Impact Polystyrene	230±10°C	50±10°C	Kapton Tape/Hairspray	Very similar to ABS but is much less likely to warp. Can be dissolved using Limonene as a solvent; great option as dissolvable support material when printing ABS with dual extrusion printer.
ASA – Acrylonitrile Styrene Acrylate	250±10°C	90±10°C	Hairspray	UV, weather, and temperature-resistant material with ABS-like properties. Best used for outdoor clips, planters, fixtures, other outdoor parts. Lighter-colored material won't yellow as quickly as ABS.
PP – PolyPropylene	250±15°C	110±10°C	Packing Tape	Elongates better than PLA, which tends to shatter and snap; PP bends and squeezes into place. Higher impact resistance. Print-surface adhesion can be difficult; Polypropylene-based packaging tape is advised. With the right tape and temperatures, it produces great prints.
Polycarbonate	290±20°C	130±15°C	GlueStick/Hairspray	Strong and very resistant to impact — used for bullet-proof glass. Extremely high temperature resistance as well. Potential uses in printed remote-control car or drone parts. An all-metal hotend for this high-temp material is required.
Conductive	220±10°C	50±10°C	Blue Painter's Tape	Slightly flexible and conductive to low power circuitry. Useful for functional prototypes, integrated circuitry, complex electronic wiring. Best printed in a dual extrusion printer, as it sticks really well to regular PLA.
Carbon Fiber Reinforced PLA	210±10°C	50±10°C	Blue Painter's Tape	Excellent structural strength and layer adhesion with very low warpage. Uses carbon fibers small enough to fit through nozzles but still provide rigidity. Ideal for items that should not bend: frames, supports, propellers, and tools. Highly abrasive; needs hardened steel nozzle for printing.

Sketch Your Ideas Thick to Thin

Use the thickness of your pens to assist in your conceptual sketching. Use a thick-line pen for broad strokes, and basic and top-level conceptualizing, and then switch to thinner and thinner pens at each stage as you refine and further detail your design. From industrial designer Reid Schlegel.

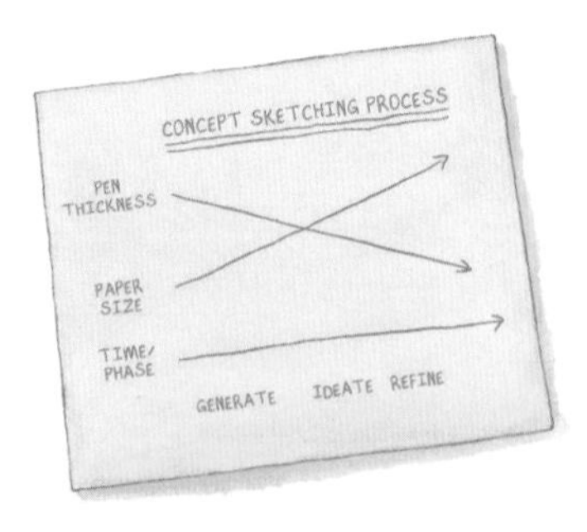

Recycling Old Electronics

MAKE SOMETHING OUT OF IT!
Before doing something suggested below, type the device name into Google, along with "DIY" or "What to do with an old...," and see if any of the projects that come up appeal to you.

GIVE IT TO A CHARITY
Goodwill is an obvious place to start (goodwillsc.org/donate/computers). The World Computer Exchange (worldcomputerexchange.org) refurbishes hardware and donates it to developing countries. Or, sell it on eBay and donate the proceeds to a charity of your choice (worldcomputerexchange.org).

SELL IT FOR CASH
Apple (apple.com/shop/trade-in) lets you trade in old hardware for new. Amazon offers a trade-in program (amzn.to/3iTOi6Q). Most manufacturers offer such programs. Check with the maker of your device.

RECYCLE IT
Find the appropriate recycler by going to call2recycle.org.

Frankenstein Prototyping

Inventor Perry Kaye has a brilliant approach to prototyping. He calls this approach "Frankenstein prototyping." When he comes up with a possible new invention, rather than drawing up plans, then paying a rapid prototyping service or someone else to fabricate it, he'd just heads to Home Depot, the Dollar Store, and the local hardware store. He finds the parts he needs on existing products (a handle here, a blade there, this motor, that gearbox). Then, he cuts up these existing products, removes the parts he needs, and cobbles them together into his new monster creation.

This is an incredibly powerful perceptual shift— to see the physical world around you as a collection of parts that are currently in one configuration but are just waiting to be taken apart and recombined into something new.

Of course, you don't need to be an inventor in the classic sense to benefit from this way of looking at the world. You can make one-off creations with this method or solve annoying design deficiencies on existing projects. We have this perceptual blindness where we tend to see things as they are rather than the potential for what they could become. Frankenstein prototyping is a way of training yourself to look for that potential.

Basic Kid Maker Skills

Here is a list of baseline skills that all well-prepared kids should be versed in before they reach adulthood. Think of it as homeschooling in self-reliance.

- Use basic scientific and critical thinking
- Perform CPR and general first aid
- Tie basic knots
- Swim
- Ride and fix a bike
- Solder and understand basic electronics
- Change a tire and oil, jumpstart a car, and change the air filter
- Generally understand combustion engines
- Cook the basics and understand food safety
- Use a fire extinguisher
- Set up a tent, build a lean-to, and collect and purify water
- Navigate with a map and compass
- Understand basic conditional logic in computer programming
- Use basic sewing techniques, including straight and whipstitch
- Safely handle fireworks, explosives, and propellants
- Fix and replace basic components in a toilet tank and shut off the water supply
- Safely handle power tools and sharpen tools and knives
- Practice basic electrical safety
- Understand house (AC) electrical systems and know where the breaker is and how to reset it
- Build a fire (Bonus points: without matches or a lighter)

Details Layer

Scott of the YouTube channel The Essential Craftsman has a wonderful statement that speaks to something very important in all aspects of making (and life): details layer. The idea here is that, in every step of a project or activity, the precision with which you do one step carries over into the next, and the next, and the next. Over time, mistakes and imperfections can compound, so it's important to do each step as well and as thoughtfully as possible.

Special Thanks

Maker's Notebook Brain Trust

R. Mark Adams
Eric Michael Beug
Boon
Terry Bronson
Jonah Brucker-Cohen
Rob Bullington
Sean Carton
Shawn Connally
Collin Cunningham
Lenore Edman
Mark Frauenfelder
Rebecca Husemann
Brian Jepson
Perry Kaye
Michelle Kempner
Jeffrey McGrew
Terrie Miller
Goli Mohammadi
Richard Nagy ("Datamancer")
Jillian Northrup
Ty Nowotny
Windell Oskay
Patti Schiendelman
Brian Scott
Jake von Slatt
Becky Stern
Jason Striegel
Phillip Torrone
Dan Woods
Natalie Zee Drieu

Acknowledgements

Fannon Printing
Alberto Gaitán
Katie Gekker
LEDtronics.com
Mister Jalopy
Ulla-Maaria Mutanen
thistothat.com
Judy Willard

When I'm working on a problem, I never think about beauty, I only think about how to solve the problem. But when I have finished, if the solution is not beautiful, I know it is wrong. — R. Buckminster Fuller

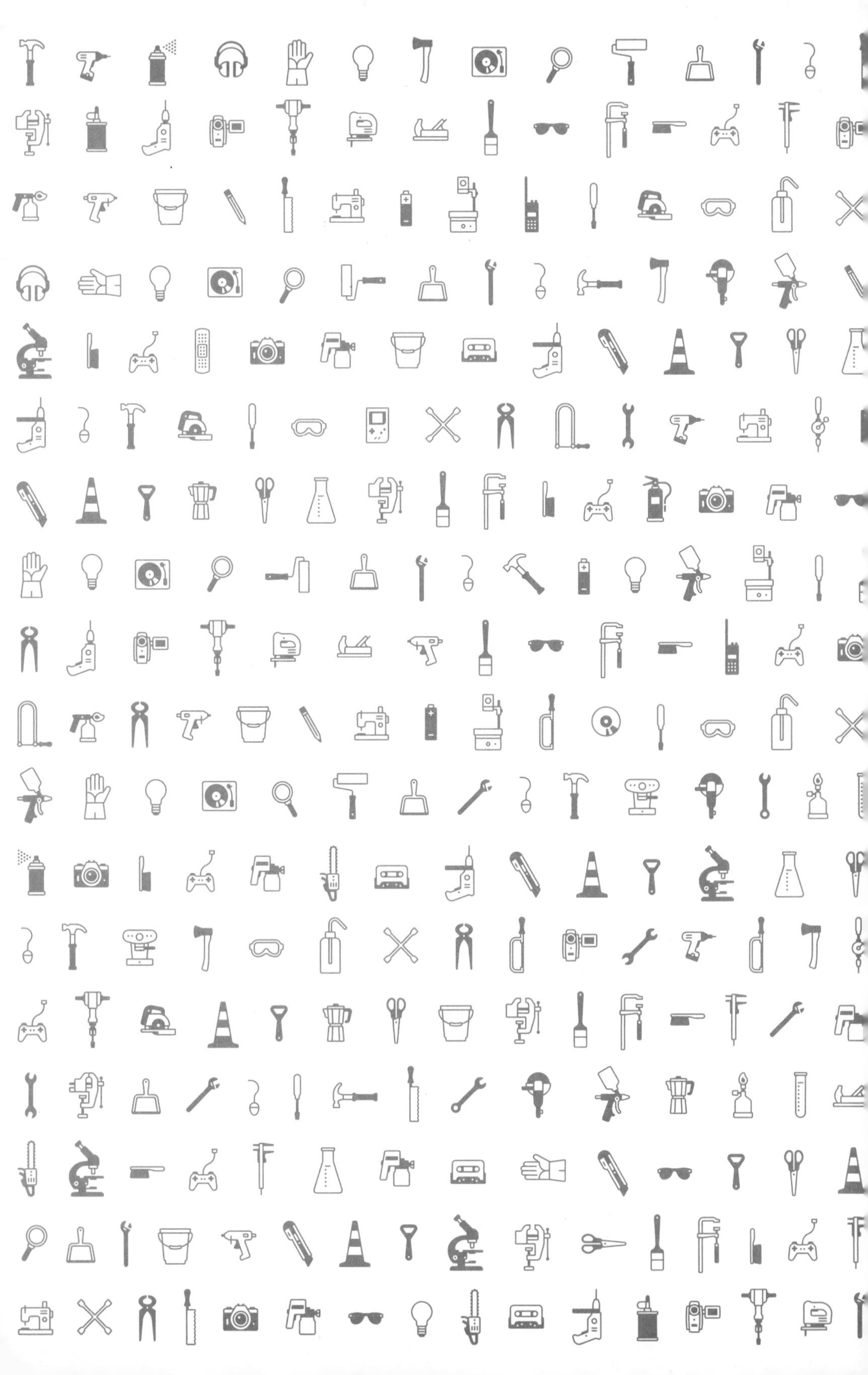